AGRICULTURAL FINANCE AND CREDIT

AGRICULTURAL FINANCE AND CREDIT

JONATHAN M. BISHOFF
EDITOR

Nova Science Publishers, Inc.
New York

NOTICE TO THE READER

The Publisher has taken reasonable care in the preparation of this book, but makes no expressed or implied warranty of any kind and assumes no responsibility for any errors or omissions. No liability is assumed for incidental or consequential damages in connection with or arising out of information contained in this book. The Publisher shall not be liable for any special, consequential, or exemplary damages resulting, in whole or in part, from the readers' use of, or reliance upon, this material.

Independent verification should be sought for any data, advice or recommendations contained in this book. In addition, no responsibility is assumed by the publisher for any injury and/or damage to persons or property arising from any methods, products, instructions, ideas or otherwise contained in this publication.

This publication is designed to provide accurate and authoritative information with regard to the subject matter covered herein. It is sold with the clear understanding that the Publisher is not engaged in rendering legal or any other professional services. If legal or any other expert assistance is required, the services of a competent person should be sought. FROM A DECLARATION OF PARTICIPANTS JOINTLY ADOPTED BY A COMMITTEE OF THE AMERICAN BAR ASSOCIATION AND A COMMITTEE OF PUBLISHERS.

LIBRARY OF CONGRESS CATALOGING-IN-PUBLICATION DATA

Agricultural finance and credit / Jonathan M. Bishoff, editor.
    p. cm.
 ISBN 978-1-60456-072-5 (hardcover)
 1. Agriculture--United States--Finance. 2. Agricultural credit--United States. 3. Agriculture--United States--Accounting. I. Bishoff, Jonathan M.
HF5686.A36A345 2008
332.7'10973--dc22
                        2007040359

Published by Nova Science Publishers, Inc. ⁺ New York

CONTENTS

PREFACE

Because of the nature of front-end funding of growing crops, cyclical weather patterns and the national security aspects of agriculture, finance and credit have become a critical component of agriculture. This new book presents important analysis dealing with issues critical to maintaining the vigorous agricultural industry in America.

Chapter 1 - The Economic Growth and Tax Relief Reconciliation Act of 2001 (EGTRRA) repeals the estate tax in 2010. During the phase-out period, the new law increases the exempt amount to $3.5 million by 2009 and lowers the top rate to 45% by 2007. The federal gift tax remains though the rate is reduced to the top personal income tax rate and the exemption is separate from the estate tax exemption. After repeal of the estate tax, carryover basis replaces step-up in basis for assets transferred at death. The legislation includes an exemption from carryover basis for capital gains of $1.3 million (and an additional $3 million for a surviving spouse). However, the estate tax provision in EGTRRA automatically sunsets December 31, 2010.

Late in the 109th Congress, two Senate compromise proposals were reported in the press, but were not introduced. One (Senator Kyl) would have set the exemption at $5 million for each spouse and lowered the rate to 15%. The second (Senator Baucus) would have set the exemption at $3.5 million and include a progressive rate schedule beginning at 15% and rising to 35%. Earlier, on July 29, 2006, the House approved H.R. 5970 by a vote of 230-180. The bill would have restored the unified estate and gift tax exclusion and raise the exclusion amount to $5 million per decedent by 2015. Any unused exclusion could have been carried over to the estate of the surviving spouse. The tax rate on taxable assets up to $25 million would have been equal to the tax rate on capital gains. The tax rate on assets over $25 million would have dropped to 30% by 2015. The JCT estimated that the estate tax provisions of H.R. 5970 would have cost $268 billion over FYs 2007-2016.

Supporters of the estate and gift tax cite its contribution to progressivity in the tax system and to the need for a tax due to the forgiveness of capital gains taxes on appreciated assets held until death. Arguments are also made that inheritances represent a windfall to heirs that are more appropriate sources of tax revenue than income earned through work and effort. Critics of the estate tax argue that it reduces savings and makes it difficult to pass on family businesses. Critics also argue that death is not an appropriate time to impose a tax; that much wealth has already been taxed through income taxes; and that complexity of the tax imposes administrative and compliance burdens that undermine the progressivity of the tax.

The analysis in this study suggests that the estate tax is highly progressive, although progressivity is undermined by avoidance mechanisms. Neither economic theory nor empirical evidence indicate that the estate tax is likely to have much effect on savings. Although some family businesses are burdened by the tax, only a small percentage of estate tax revenues are derived from family businesses. Even though there are many estate tax avoidance techniques, it also is possible to reform the tax and reduce these complexities as an alternative to eliminating the tax. Thus, the evaluation of the estate tax may largely turn on the appropriateness of such a revenue source and its interaction with incentives for charitable giving, state estate taxes, and capital gains and other income taxes. This report will be updated as legislative events warrant.

Chapter 2 - The 2001 tax revision began a phaseout of the estate tax, by increasing exemptions and lowering rates. The estate tax is scheduled to be repealed in 2010 and a provision to tax appreciation on inherited assets (in excess of a limit) will be substituted. The 2001 tax provisions sunset, however, so that absent a change making them permanent the estate tax will revert, in 2011, to prior, pre-2001, law. Proposals to make the repeal permanent, or to significantly increase the exemptions and lower the rate, are under consideration.

Currently, discussions of the estate tax are focusing particular attention on the effects on family businesses, including farms, and perception that the estate tax unfairly burdens family businesses because much of the estate value is held in illiquid assets (e.g., land, buildings, and equipment). The estate tax may even force the liquidation of family businesses. A special family business deduction, the Qualified Family Owned Business Interest Exemption (QFOBI) was enacted in 1997. Presently, because of higher exemptions allowed and a previous cap on the combined regular and small business exemption, this provision is no longer relevant. If, however, the estate tax repeal sunsets, QFOBI will again be germane. In the 109th Congress, H.R. 8, which would make the estate tax repeal permanent, was passed by the House, but not by the Senate. There were also proposals to allow an expanded business exemption (H.R. 1612 and S. 928) as well as proposals to allow a higher exemption (H.R. 1577 and H.R. 1574) or both a higher exemption and lower rate (H.R. 1560, H.R. 1568, H.R. 1614, and H.R. 5638). H.R. 5970 — a proposal for a credit eventually equivalent to a $5 million exemption ($10 million for a married couple) with tax rates initially set at the capital gains tax rate (currently 15%, and scheduled to rise to 20%) for estates up to $25 million, and at twice the gains rate for those over $25 million — was passed by the House on July 29, 2006.

Evidence suggests, however, that only a small fraction of estates with small or family business interests have paid the estate tax (about 3.5% for businesses in general, and 5% for farmers, compared to 2% for all estates). Recent estimates suggest that only a tiny fraction of family-owned businesses (less than ½ of 1%) are subject to the estate tax but do not have readily available resources to pay the tax. Thus, while the estate tax may be a burden on those families, the problem is confined to a small group.

If the estate tax is repealed, QFOBI will allow an exemption for some or all of business assets in about a third to a half of estates with more than half their assets in these businesses, but the value of the exemption will be reduced because the general exemption has increased. If the estate tax repeal is made permanent, liquidity will cease being a problem, although family businesses may be more likely than other estates to be affected by the capital gains provisions. Exposure to the estate tax, if it is reinstated, would be significantly decreased by increases in either the family business or general exemptions. The report also discusses an uncapped exemption and an uncapped exemption targeted at liquidity issues. This report will be updated as legislative events warrant.

Chapter 3 - This report provides a basic explanation of how to calculate the federal estate tax liability for a taxable estate of any given size, using the schedule of marginal tax rates and the applicable exclusion amount or the applicable credit amount for the year of death. The applicable exclusion amount is the amount of any decedent's taxable estate that is free from tax, known informally as the estate tax exemption. The applicable credit amount is the corresponding tax credit equal to the tax that would be due on a taxable estate the size of the applicable exclusion amount.

A shortcut is available to calculate the tax on the estates of decedents dying in 2006 through 2009. The estate tax liability can be calculated simply by multiplying the amount of the taxable estate in excess of the applicable exclusion amount for the year of death times the maximum estate tax rate for the year. The applicable exclusion amount is $2 million for 2006-2008 and $3.5 million for 2009. The maximum tax rate is 46% for 2006 and 45% for 2007-2009.

A more formal method is required to calculate the tax liability for years before 2006 or for 2011 and beyond. This is because more than one marginal tax rate applies to taxable estate values in excess of the exclusion amount. First, the tentative tax that would be due on the entire taxable estate is calculated from the marginal tax rate table for the year of death. Then, the applicable credit amount for the year of death is subtracted from the tentative tax to determine the tax due.

A numerical example is presented in the text and in worksheets for a $5 million taxable estate of a decedent dying in 2007 or 2008. Both the shortcut and formal methods are used to calculate the tax liability.

The Economic Growth and Tax Relief Reconciliation Act of 2001 (P.L. 107-16, EGTRRA) gradually lowered the maximum estate tax rate and substantially raised the applicable exclusion amount over the years 2002 through 2009. The maximum tax rate fell from 60% under prior law in 2001 (a 55% marginal rate on taxable estate values over $3 million plus a 5% surtax from $10 million to $17 million) to 45% in 2007-2009. The applicable exclusion amount rose from $675,000 in 2001, in steps, up to $3.5 million in 2009. EGTRRA repealed the estate tax for decedents dying in 2010. However, all of the provisions under the act are scheduled to sunset on December 31, 2010. Unless changed beforehand, in 2011 the law will revert back to what it would have been had EGTRRA never been enacted. Under a provision of the Taxpayer Relief Act of 1997 (P.L. 105-34), in 2011 and beyond the applicable exclusion amount would be $1 million, in contrast to $675,000 in 2001. The tables in the appendix present the marginal estate tax rates and the applicable exclusion amount for each year. This report will be updated when there are changes in the law governing estate taxes.

Chapter 4 - Tax legislation in 1997 reduced capital gains taxes on several types of assets, imposing a 20% maximum tax rate on long-term gains, a rate temporarily reduced to 15% for 2003-2008, which was extended for two additional years in 2006. There is also an exclusion of $500,000 ($250,000 for single returns) for gains on home sales. The capital gains tax has been a tax cut target since the 1986 Tax Reform Act treated capital gains as ordinary income. An argument for lower capital gains taxes is reduction of the lock-in effect. Some also believe that lower capital gains taxes will cost little compared to the benefits they bring and that lower taxes induce additional economic growth, although the magnitude of these potential effects is in some dispute. Others criticize lower capital gains taxes as benefitting higher income individuals and express concerns about the budget effects, particularly in future years. Another criticism of lower rates is the possible role of a larger capital gains tax differential in encouraging tax sheltering activities and adding complexity to the tax law.

Chapter 5 - The federal government has a long history of providing credit assistance to farmers by issuing direct loans and guarantees, and creating rural lending institutions. These institutions include the Farm Service Agency (FSA) of the U.S. Department of Agriculture (USDA), which makes or guarantees loans to farmers who cannot qualify at other lenders, and the Farm Credit System (FCS), which is a network of borrower-owned lending institutions operating as a government-sponsored enterprise.

The 110th Congress is expected to address agricultural credit through both appropriations and authorizations bills. Appropriators will consider funding for FSA's farm loan programs, and the agriculture committees may consider changes to FSA and FCS lending programs. The 2007 farm bill is expected to be the venue for many of the authorizing issues, although stand-alone legislation may be used for extensive reforms. This report will be updated.

Chapter 6 - In an unprecedented move, an institution of the Farm Credit System (FCS) — a government-sponsored enterprise — initiated procedures on July 30, 2004, to leave the FCS and be purchased by a private company. But after much controversy, including congressional hearings, the board of directors of Farm Credit Services of America (FCSA) voted on October 19, 2004, to terminate its agreement with Rabobank before seeking approval from the Farm Credit Administration, the System's federal regulator.

FCSA is the FCS lending association serving Iowa, Nebraska, South Dakota, and Wyoming. Rabobank is a private Dutch banking company with extensive experience in agriculture and a growing global network. Under the plan, the loans, facilities, and employees of FCSA would have become part of Rabobank, and new FCS charters would have been issued to reestablish a System presence in the four-state region.

The option to leave the System is allowed by statute under the Farm Credit Act of 1971, as amended, but has been exercised only once, and did not involve an outside purchaser. Although Congress had no direct statutory role in the approval process, the House held hearings on the implications of the deal, and Senators Daschle and Johnson introduced S. 2851 to require public hearings and a longer approval process. This report will not be updated.

Chapter 7 - The Farm Credit System (FCS) is a nationwide financial cooperative lending to agricultural and aquatic producers, rural homeowners, and certain agriculture-related businesses and cooperatives. Established in 1916, this government-sponsored enterprise (GSE) has a statutory mandate to serve agriculture. It receives tax benefits, but no federal appropriations or guarantees. FCS is the only direct lender among the GSEs. Farmer Mac, a separate GSE but regulated under the umbrella of FCS, is a secondary market for farm loans. Federal oversight by the Farm Credit Administration (FCA) provides for the safety and soundness of FCS institutions.

Current issues and legislation affecting the FCS are discussed in CRS Report RS21977, *Agricultural Credit: Institutions and Issues.* This report will be updated.

Chapter 8 - Chapter 12 of the U.S. Bankruptcy Code dealing with "family farmer" reorganization, temporarily extended 11 times since its original enactment, is made permanent by enactment of the Bankruptcy Abuse Prevention and Consumer Protection Act,

P.L. 109-8. It is amended to include "family fisherman" as well. This report surveys the highlights of this chapter.

Chapter 9 - Supporters of the estate and gift tax argue that it provides progressivity in the federal tax system, provides a backstop to the individual income tax, and appropriately targets assets that are bestowed on heirs rather than assets earned through their hard work and effort. Progressivity, however, can be obtained through the income tax and the estate and gift tax is an imperfect backstop to the income tax. Critics argue that the tax discourages savings, harms small businesses and farms, taxes resources already subject to income taxes, and adds to the complexity of the tax system. Critics also suggest death is an inappropriate time to impose a tax. The effect on savings, however, is uncertain, most farms and small businesses do not pay the tax, and complexity could be reduced through reform of the tax. This report will be updated as legislative developments warrant.

In: Agricultural Finance and Credit
Editor: J. M. Bishoff, pp. 1-29

Chapter 1

ESTATE AND GIFT TAXES: ECONOMIC ISSUES*

Jane G. Gravelle[1] and Steven Maguire[2]

[1] Senior Specialist in Economic Policy Government and Finance Division
[2] Analyst in Public Finance Government and Finance Division

ABSTRACT

The Economic Growth and Tax Relief Reconciliation Act of 2001 (EGTRRA) repeals the estate tax in 2010. During the phase-out period, the new law increases the exempt amount to $3.5 million by 2009 and lowers the top rate to 45% by 2007. The federal gift tax remains though the rate is reduced to the top personal income tax rate and the exemption is separate from the estate tax exemption. After repeal of the estate tax, carryover basis replaces step-up in basis for assets transferred at death. The legislation includes an exemption from carryover basis for capital gains of $1.3 million (and an additional $3 million for a surviving spouse). However, the estate tax provision in EGTRRA automatically sunsets December 31, 2010.

Late in the 109th Congress, two Senate compromise proposals were reported in the press, but were not introduced. One (Senator Kyl) would have set the exemption at $5 million for each spouse and lowered the rate to 15%. The second (Senator Baucus) would have set the exemption at $3.5 million and include a progressive rate schedule beginning at 15% and rising to 35%. Earlier, on July 29, 2006, the House approved H.R. 5970 by a vote of 230-180. The bill would have restored the unified estate and gift tax exclusion and raise the exclusion amount to $5 million per decedent by 2015. Any unused exclusion could have been carried over to the estate of the surviving spouse. The tax rate on taxable assets up to $25 million would have been equal to the tax rate on capital gains. The tax rate on assets over $25 million would have dropped to 30% by 2015. The JCT estimated that the estate tax provisions of H.R. 5970 would have cost $268 billion over FYs 2007-2016.

Supporters of the estate and gift tax cite its contribution to progressivity in the tax system and to the need for a tax due to the forgiveness of capital gains taxes on appreciated assets held until death. Arguments are also made that inheritances represent a windfall to heirs that are more appropriate sources of tax revenue than income earned through work and effort. Critics of the estate tax argue that it reduces savings and makes

---

* Excerpted from CRS Report RL30600, dated January 26, 2007.

it difficult to pass on family businesses. Critics also argue that death is not an appropriate time to impose a tax; that much wealth has already been taxed through income taxes; and that complexity of the tax imposes administrative and compliance burdens that undermine the progressivity of the tax.

The analysis in this study suggests that the estate tax is highly progressive, although progressivity is undermined by avoidance mechanisms. Neither economic theory nor empirical evidence indicate that the estate tax is likely to have much effect on savings. Although some family businesses are burdened by the tax, only a small percentage of estate tax revenues are derived from family businesses. Even though there are many estate tax avoidance techniques, it also is possible to reform the tax and reduce these complexities as an alternative to eliminating the tax. Thus, the evaluation of the estate tax may largely turn on the appropriateness of such a revenue source and its interaction with incentives for charitable giving, state estate taxes, and capital gains and other income taxes. This report will be updated as legislative events warrant.

The estate and gift tax has been and will continue to be the subject of significant legislative interest. The "Economic Growth and Tax Relief Reconciliation Act of 2001" (EGTRRA, P.L. 107-16) repeals the estate tax after 2009. However, the legislation sunsets after 2010 reverting back to the law as it existed in 2001. Congress could eliminate the sunset provision in EGTRRA, thus making repeal of the estate tax permanent. Repeal of the sunset would retain the EGTRRA changes to the taxation of capital gains of inherited assets and the gift tax.

Immediate repeal of the estate and gift tax in 2001 would have cost up to $662 billion (over 10 years), an amount in excess of the projected estate tax yield of $409 billion because of projected behavioral responses that would also lower income tax revenues (e.g. more life time transfers to donees in lower tax brackets, more purchase of life insurance with deferral aspects, and lower compliance). Repealing the EGTRRA sunset would cost $290.0 billion over the 2006-2015 budget window. Most of the revenue loss is in the "out" years; $9.1 billion over 2006 to 2010 and $280.9 billion over 2011 to 2015.[1] Immediate repeal (as opposed to eventual repeal beyond 2010) would be more expensive.

Toward the end of the 109th Congress, two compromise proposals were reported in the press, but not formally introduced. One, offered by Senator Kyl, would have set the exemption at $5 million for each spouse ($10 million per couple) and lowered the tax rate to 15%. The second, offered by Senator Baucus, would have set the exemption at $3.5 million and include a progressive rate schedule beginning at 15% and rising with the size of the estate to 35%.[2] One independent estimate of the cost of Senator Kyl's proposal projected that it would have cost approximately 84% of full repeal.[3] The relatively low rate of 15% is responsible for most of the revenue loss under the Kyl proposal. No potential revenue loss estimates are available for the Baucus proposal, although it would be less expensive than the Kyl proposal.

Proponents of an estate and gift tax argue that it contributes to progressivity in the tax system by "taxing the rich." (Note, however, that there is no way to objectively determine the optimal degree of progressivity in a tax system). A related argument is that the tax reduces the concentration of wealth and its perceived adverse consequences for society.[4] Moreover, while the estate and gift tax is relatively small as a revenue source (yielding $24.7 billion in 2005 and accounting for 1.2% of federal revenue), it raises a not insignificant amount of revenue — revenue that could increase in the future due to strong performance of stock market and growth in intergenerational transfers as the baby boom generation ages. Eliminating or reducing the tax would either require some other tax to be increased, some spending program to be reduced, or an increase in the national debt.

In addition, to the extent that inherited wealth is seen as windfall to the recipient, such a tax may be seen by some as fairer than taxing earnings that are the result of work and effort. Finally, many economists suggest that an important rationale for maintaining an estate tax is the escape of unrealized capital gains from any taxation, since heirs receive a stepped-up basis of assets. Families that accrue large gains through the appreciation of their wealth in assets can, in the absence of an estate tax, largely escape any taxes on these gains by passing on the assets to their heirs.

The estate tax also encourages giving to charity, since charitable contributions are deductible from the estate tax base. Since charitable giving is generally recognized as an appropriate object of subsidy, the presence of an estate tax with such a deduction may be seen as one of the potential tools for encouraging charitable giving.

Critics of the estate and gift tax typically make two major arguments: the estate and gift tax discourages savings and investments, and the tax imposes an undue burden on closely held family businesses (including farms). In the latter case, the argument is made that the estate tax forces the break-up of family businesses without adequate liquidity to pay the tax. Critics also suggest that the estate and gift tax is flawed as a method of introducing progressivity because there are many methods of avoiding the tax, methods that are more available to very wealthy families (although this criticism could support reform of the tax as well as repeal). A related criticism is that the administrative and compliance cost of an estate and gift tax is onerous relative to its yield (again, however, this argument could also be advanced to support reform rather than repeal). In general, there may also be a feeling that death is not a desirable time to impose a tax; indeed, the critics of the estate and gift tax often refer to the tax as a death tax. Critics also argue that some of the wealth passed on in estates has generally already been subject to capital income taxes.

The remainder of this report, following a brief explanation of how the tax operates, analyzes these arguments for and against the tax. The report concludes with an inventory and discussion of alternative policy options.

HOW THE ESTATE AND GIFT TAX WORKS

The unified estate and gift tax is levied on the transfer of assets that occurs when someone dies or gives a gift. Filing an estate tax return can be difficult depending on the value and complexity of the estate. The purpose here is to outline the mechanics of the estate and gift tax. The first section begins with a brief review of the general rules accompanied with a numerical example. There are some minor provisions of the law that are not discussed here, however, such as the phase out of the graduated rates and the credit for taxes on property recently transferred.[5] The second section summarizes the special rules for farms and small businesses. And, the final section briefly describes the generation skipping transfer tax. The appendix of this report provides detailed data from returns filed in 2005, the latest year for which data are available.

General Rules

Filing Threshold

In 2007, estates valued over $2.0 million must file an estate tax return. The applicable credit, which is identical to the filing threshold, effectively exempts from taxation the portion of the estate that falls below the filing threshold. (The filing threshold is lower, however, if gifts have already been made.) Table A1 in the appendix reports the current filing requirement and the unified credit equivalent for 2004 through 2011.

Gross Estate Value

The gross estate value, which was $178.1 billion for returns filed in 2005, is the total value of all property and assets owned by decedents. Table A2 in the appendix provides the gross estate value for the returns filed in 2005 by wealth category. The data represent the returns filed in 2005, not the decedents in that year. Thus, a portion of the returns filed in 2005 are from estates valued in years before 2005.

Allowable Deductions

Deductions from the estate reduce the taxable portion of the gross estate and in turn the number of taxable returns. In 2005, $83.1 billion was deducted from estates. The most valuable deduction is for bequests to a surviving spouse, $53.7 billion; the most prevalent (though smallest reported) deduction is for funeral expenses, $326.4 million. Appendix table A3 lists the deductions in greater detail for returns filed in 2005. Beginning in 2005, estates may deduct state estate taxes paid. Before 2005, taxpayers received a federal credit for state death taxes paid. That credit was phased out incrementally from 2002 through 2004.

Taxable Estate

After subtracting allowable deductions, the remainder of the estate is the taxable estate. Taxable estate value was $91.5 billion in 2005. Adjusted taxable gifts are then added to the taxable estate to arrive upon the *adjusted* taxable estate. An individual is allowed to exclude $12,000 in gifts per year per donee from taxable gifts. Thus, only the amount exceeding the $12,000 limit is added back to the taxable estate. Only 8,589 returns filed in 2005 included taxable gifts, adding approximately $6.3 billion to the total estate value. Thus, adjusted taxable estates were worth $101.3 billion in 2005. Generally, the adjusted taxable estate represents the base of estate tax.

Rates and Brackets

After establishing the value of the taxable estate, the executor calculates the tentative estate tax due.[6] The tax due is tentative because the executor has not redeemed the *applicable credit amount*.[7] As noted earlier, the credit for state estate and inheritance taxes was repealed and replaced with a deduction beginning in 2005.

A Numerical Example

The remaining steps in calculating the estate and gift tax are most easily exhibited through numerical example. To accomplish this, we first assume a decedent, who dies in 2007, has an estate worth $5 million and leaves $2 million to his wife and contributes $400,000 to a charitable organization. We also assume the decedent has not made any taxable gifts leaving $2.6 million in his estate after deductions. This simple example is exhibited below.

Table 1. Numerical Example
(2007 rules)

Gross Estate Value	$5,000,000
— Less: hypothetical marital deduction	$2,000,000
— Less: hypothetical charitable contribution deduction	$400,000
Taxable Estate	$2,600,000

The taxable estate is valued at $2.6 million after the allowable deductions have been subtracted from the gross estate value.[8] The tax is applied to the $2.6 million in increments of estate value as provided for in the tax code. For example, the first increment of $10,000 is taxed at 18%, the second increment of $10,000 is taxed at 20%, the third increment of $20,000 is taxed at 22%, etc. This process continues until the entire $2.6 million is taxed. The last increment of estate value, that from $1.5 million to $2.6 million, is taxed at a 45% rate. Thus, even though this estate is in the 45% bracket, only a portion ($1.1 million) of the estate is taxed at the 45% rate.

Tentative Estate Tax
In 2005, the aggregate tentative estate tax after deductions and before credits was $42.7 billion. Returning to our example, the $2.6 million taxable estate yields a tentative estate tax of $1,050,800. Recall, however, we have not yet considered the "applicable credit."

The Applicable Credit (Unified with Gift Tax before 2004)
For decedents dying in 2007, the applicable credit is $780,800, which leaves an estate tax due in our example of $270,000. The applicable credit reduced the tentative estate tax by $20.6 billion in 2005.

Federal Credit for State Death Taxes (Eliminated in 2005)
The state death tax credit reduced the federal estate tax due by $1.8 billion in 2005.[9] This tax credit is determined by yet another tax rate schedule. The taxable estate value, which is $2.6 million in our example, is reduced by a standard exemption of $60,000 and the credit rate schedule applies to the remainder. EGTRRA reduces and eventually repeals the credit for state death taxes. In 2004, the credit was 25% of what the credit would have been before EGTRRA. In 2005, the credit was repealed and estates were allowed to deduct state death taxes paid. Table A5 of the appendix reproduces the now-repealed credit schedule for state death taxes. For our hypothetical estate filed in 2007, the credit is not available and thus state death taxes would be deductible. In many states, however, the state estate tax is repealed along with the federal credit.

Net Federal Estate Tax

The net estate tax due was $21.5 billion in 2004.[10] This is the final step for the estate executor. After all exemptions, deductions, and credits, the $5 million dollar estate we began with must now remit $270,000 to the federal government.

All of the steps described above are included in table 2. Also, an estimate of the average estate tax rate is presented in the bottom row. The federal rate is calculated as the federal estate tax due divided by the gross estate value.

Table 2. Numerical Example Continued with Taxes and Credits (2007 rules)

Gross Estate Value	$5,000,000
— Less: hypothetical marital deduction	$2,000,000
— Less: hypothetical charitable contribution deduction	$400,000
Taxable Estate	$2,600,000
Tentative Estate Tax (from the rate schedule)	$1,050,800
— Less: Applicable Credit Amount (in 2007)	$780,800
Net Federal Estate Tax	$270,000
Average Effective Federal Estate Tax Rate	5.40%

Special Rules for Family Owned Farms and Businesses

There are primarily two special rules for family owned farms and businesses. The first special rule (26 I.R.C. 6166) allows family owned farm and business estates to pay the tax in installments over a maximum of 10 years after a deferment of up to five years. The farm or business must comprise at least 35% of the adjusted gross estate value to qualify for the installment method. A portion of the deferred estate tax is assessed an annual 2% interest charge.

The second special rule (26 I.R.C. 2032A) allows family farms and businesses that meet certain requirements to value their land as currently used rather than at fair market value. To avoid a recapture tax, heirs must continue to use the land as designated in the special use notice for at least 10 years following the transfer. The market value can be reduced by a maximum $750,000 in 1998. After 1998, the maximum is indexed for inflation, rounded to the next lowest multiple of $10,000. In 2006, the maximum was $900,000.

The Generation Skipping Transfer Tax

Generally, the generation skipping transfer (GST) tax is levied on transfers from the decedent to grandchildren. The tax includes a $2,000,000 exemption per donor in 2007 that is pegged to the general estate tax exclusion. Married couples are allowed to "split" their gifts for an effective exemption of $3,000,000. The rate of tax is the highest estate and gift tax rate or 45% in 2007. These transfers are also subject to applicable estate and gift taxes. The GST exemption rises to $3.5 million in 2009. Very few estates pay a generation skipping transfer tax because the high rate of tax discourages this type of bequest.

ECONOMIC ISSUES

As noted in the introduction, the principal arguments surrounding the estate and gift tax are associated with the desirability of reducing the concentration of wealth and income through the tax and the possible adverse effect of the tax on savings behavior and family businesses. There are a number of other issues of fairness or efficiency associated with particular aspects of the tax (e.g. marital deductions, charitable deductions, effects on small businesses, interaction with capital gains taxes) , and the possible contribution to tax complexity. These issues are addressed in this section.

The Distributional Effect of the Estate and Gift Tax

Distributional effects concern both vertical equity (how high income individuals are affected relative to low income individuals) and horizontal equity (how individuals in equal circumstances are differentially affected). Note that economic analysis cannot be used to determine the optimal degree of distribution across income and wealth (vertical equity).

Vertical Equity
The estate tax is the most progressive of any of the federal taxes; out of the approximately 2.4 million deaths in 2004, only 1.3% of estates paid any estate tax.[11] This percentage can be contrasted with the income tax where most families and single individuals file tax returns and about 70% of those returns owe tax. In addition, out of the 1.3% of decedents whose estates pay tax, about 70% of these had gross estates valued between $1 million and $2.5 million in 2004, which are the smallest (based on gross estate value) taxable estates.

Evidence suggests that the average effective tax rate rises with the size of the estate except for the highest tax rate bracket, as shown in table 1 [columns (f) and (g)]. Column (f) reports 2005 effective tax rates for the decedent before the credit for state death taxes and column (g) shows the actual amount paid to the federal government after all credits. Estates valued at less than the exemption amount, of course, pay no taxes and the tax rate rises and then falls with the very largest estates, despite the fact that the rates are graduated.

Columns (b), (c), and (d) show the deductions from the estate as a percentage of gross estate value. Charitable deductions are the primary reason for the lower tax rate in the highest levels of the estate tax. The charitable deduction accounts for 11.0% of estates on average but 24.3% in the highest wealth bracket. The deduction for bequests left to spouse also rises as a portion of the gross estate as estate size increases. The progressivity of the estate tax for the estates valued at less than $10 million is the result of the unified credit and the graduated rate structure.

The data in table 3 may actually overstate the amount of rate progression in the estate tax. Tax planning techniques, such as gift tax exclusions or valuation discounts, reduce the size of the gross estate and are more common with larger estates. These techniques reduce the size of the estate but do not appear in the IRS data, thus, the effective tax rates may be overstated for larger estates.

Despite the lack of progressivity through all of the estate size brackets, the principal point for distributional purposes is that the estate and gift tax is confined to the wealthiest of decedents and to a tiny share of the population. For example, estates over $5 million accounted for 19.5% of taxable estates, but accounted for 63.0% of estate tax revenues in 2005. Thus, to the extent that concentration of income and wealth are viewed as undesirable, the estate tax plays some role, albeit small — because few pay the tax — in increasing income and wealth equality.

Note also an effect that contradicts some claims made by opponents of the tax. The Carnegie conjecture suggests that large inheritances reduce labor effort by heirs.[12] Thus, the estate tax, which reduces inheritances, could increase output and economic growth because heirs work more (increase their labor supply) if their inheritance is reduced. Although, for very large inheritances, the effect of one individual on the labor supply may be small relative to the effect on saving.

Table 3. Estate Tax Deductions and Burdens, 2005

Size of Gross Estate ($ millions)	Percent of Gross Estate			Tax as a Percent of Net Estate[a]		
	Expenses	Bequests to Spouse	Charity	Before Credit	After Unified Credit	After All Credits[b]
(a)	(b)	(c)	(d)	(e)	(f)	(g)
1.5-2.5	5.15%	19.77%	3.93%	28.89%	-0.28%	4.04%
2.5-5.0	5.97%	28.66%	5.14%	27.07%	10.61%	11.64%
5.0-10.0	6.46%	34.09%	7.03%	25.91%	17.76%	17.09%
10.0-20.0	5.85%	38.31%	8.51%	24.05%	20.06%	18.66%
over 20.0	4.39%	34.41%	24.30%	18.53%	17.73%	16.10%
Total	5.40%	30.14%	10.99%	24.53%	12.28%	12.77%

Source: CRS calculations from Statistics of Income, *Estate Tax Returns Filed in 2005*, IRS, SOI unpublished data, November 2006.
a. Net estate is estate value less expenses. Expenses include funeral expenses, attorney's fees, executors' commissions, other expenses/losses, and debts and mortgages. b. This includes any gift taxes that are owed by an estate, which could increase the total taxes owed by an estate.

Horizontal Equity

Estate and gift taxes can affect similar individuals differentially for a variety of reasons. Special provisions for farmers and family businesses (discussed subsequently) can cause families with the same amount of wealth to be taxed differentially. The availability and differential use of avoidance techniques (also discussed subsequently) can lead to different tax burdens for the same amount of wealth. Moreover, individuals who accumulate similar amounts of wealth may pay differential taxes depending on how long they live.

Effect on Saving

Many people presume that the estate tax reduces savings, since the estate and gift tax, like a capital income tax, applies to wealth. It may appear "obvious" that a tax on wealth would reduce wealth. However, taxes on capital income do not necessarily reduce savings. This ambiguous result arises from the opposing forces of an income and substitution effect. An investment is made to provide future consumption; if the rate of return rises because a tax is cut, more consumption might be shifted from the present to the future (the substitution effect). This effect, in isolation, would increase saving.

However, the tax savings also increases the return earned on investment and allows higher consumption both today and in the future. This effect is called an income effect, and it tends to reduce saving. Its effect is most pronounced when the savings is for a fixed target (such as a fund for college tuition or a target bequest to an heir). Thus, saving for precautionary reasons (as a hedge against bad events) is less likely to increase when the rate of return rises than saving for retirement. Empirical evidence on savings responses, while difficult to obtain, suggests a small effect of uncertain sign (i.e. either positive or negative). Current events certainly suggest that savings fall when the rate of return rises: as returns on stocks have increased dramatically, the savings rate has plunged.

The same points can generally be made about a tax on estates and gifts, although some analysts suspect that an estate tax, to be paid at a distant date in the future, would be less likely to have an effect (in either direction) than income taxes being paid currently. A reduction in estate taxes makes a larger net bequest possible, reducing the price of the bequest in terms of forgone consumption. This substitution effect would cause savings to increase. At the same time, a reduction in estate taxes causes the net estate to be larger, allowing a larger net bequest to be made with a smaller amount of savings (the income effect). Again, the latter effect is most pronounced when there is a target net bequest; a smaller gross bequest can be left (and less savings required on the part of the decedent) to achieve the net target.

Unfortunately, virtually no empirical evidence about the effect of estate and gift taxes exists, in part because these taxes have been viewed as small and relatively unimportant by most researchers and in part because there are tremendous difficulties in trying to link an estate and gift tax which occurs at the end of a lifetime to annual savings behavior. But a reasonable expectation is that the effects of cutting the estate and gift tax on savings would not be large and would not even necessarily be positive.

Of course, the effect on national saving depends on the use to which tax revenues are put. If revenues are used to decrease the national debt, they become part of government saving, and it is more likely that cutting estate and gift taxes would reduce saving by decreasing government saving, since there may be little or no effect on private saving. If they are used for government spending on consumption programs, or transfers that are primarily used for consumption, then it is less likely that cutting estate and gift taxes would reduce saving because the estate tax cut would be financed out of decreased consumption (rather than decreased saving). In this case, reducing the tax would probably have a small effect on national saving, since the evidence suggests a small effect on private saving. A similar effect would occur if tax revenues are held constant and the alternative tax primarily reduced consumption.

Actually, the estate and gift tax is, in some ways, more complicated to assess than a tax on capital income or wealth. There are a variety of possible motives for leaving bequests, which are likely to cause savings to respond differently to the estate tax. In addition, there are consequences for the heirs which may affect their savings.

Several of these alternative motives and their consequences are outlined by Gale and Perozek.[13] Motives for leaving bequests include (1) altruism: individuals want to increase the welfare of their children and other descendants because they care about them; (2) accident: individuals do not intentionally save to leave a bequest but as a fund to cover unexpected costs or the costs of living longer than expected (thus, bequests are left by accident and are in the nature of precautionary savings); (3) exchange: parents promise to leave bequests to their children in exchange for services (visiting, looking after parents when they are sick); and (4) joy of giving: individuals get pleasure directly from giving, with the pleasure depending on the size of the estate. To Gale and Perozek's classifications we might add satiation: when individuals have so much wealth that any consumption desire can be met.

The theoretical effects of these alternative theories on decedents and heirs are summarized in table 4. A discussion of each follows in the text, but it is interesting to see that there is a tendency for estate taxes to increase saving, not decrease it. This effect occurs in part because there are "double" income effects that discourage consumption, acting on both the decedent and the heir.

Altruistic

When giving is motivated by altruism, the effect of the tax is ambiguous, as might not be surprising given the discussion of income and substitution effects. The effects on the parents are ambiguous, while the windfall receipt of an inheritance tends to reduce the need to save by the children. That is, the estate tax reduces the inheritances and thus increases saving by heirs. The outcomes are also partly dependent on whether children think they can elicit a larger inheritance by squandering their own money (which causes them to save even less) and whether the parent sees this problem and responds to it in a way that forestalls it. Interestingly, some parents might respond by spending a lot of their assets before death to induce their children to be more responsible and save more. The cost of doing this is the reduction in welfare of their children from the smaller bequest as compared with the parent's benefit from consumption. The estate tax actually makes the cost of using this method smaller (in terms of reduced bequests for each dollar spent), and causes the parents to consume more. While these motivations and actions of parent and child can become complex, this theory leaves us with an ambiguous effect on savings.

Accidental

In the second case, where bequests are left because parents die before they have exhausted their resources, the estate tax has no effect on the saving of the parents. Indeed, the parents are not really concerned about the estate tax since it has no effect on the reason they are accumulating assets. If they need the assets because they live too long or become ill, no tax will be paid. Bequests are a windfall to children, in this case, and tend to increase their consumption. Thus, taxing bequests, because it reduces this windfall, reduces their consumption and promotes savings. If the revenue from the estate tax is saved by the government, national saving rises. (If the revenue is spent on consumption, there is no effect on savings.) Thus, in this case, the estate tax reduces private consumption and repealing it, reducing the surplus, would increase consumption (reduce savings).

Exchange

In the third case, parents are basically paying for children's services with bequests and the estate tax becomes like a tax on products: the price for their children's attention has increased. Not surprisingly, the savings and size of bequest by the parents depends on how responsive they are to these price changes. If the demand is less responsive to price changes (price inelastic), parents will save and bequeath more to make up for the tax to be sure of receiving their children's services, but if there are close substitutes they might save less, bequeath less, and purchase alternatives (e.g. nursing home care). In this model, the child's saving is not affected, since the bequest is payment for forgone wages (or leisure).

Table 4. Theoretical Effect of Estate Tax on Saving, By Bequest Motive

Bequest Motive	Effect on Decedent Saving	Effect on Heir Saving
Altruism	Ambiguous	Increases
Accidental	None	Increases
Exchange	Ambiguous	None
Joy of Giving	Ambiguous	Increases
Satiation	None	Increases or None

Joy of Giving

A fourth motive is called the "joy-of-giving" motive, where individuals simply enjoy leaving a bequest. If the parent focuses on the before-tax bequest, the estate tax will have no effect on his or her behavior, but will reduce the inheritance and theoretically increase the saving of children. Thus, repealing the estate tax would reduce private saving. If the parent focuses on the after-tax bequest, the effect on saving is ambiguous (again, due to income and substitution effects).

Satiation

Some families may be so wealthy that they can satisfy all of their consumption needs without feeling any constraints and their wealth accumulation may be a (large) residual. In this case, as well, the estate tax would have no effect on saving of the donor, and perhaps little effect on the donee as well.

Empirical Evidence

Evidence for these motives is not clear but this analysis does suggest that there are many circumstances in which a repeal of the estate tax would reduce savings, not increase it. Virtually no work has been done to estimate the effect of estate taxes on accumulation of assets. A preliminary analysis of estate tax data by Kopczuk and Slemrod found some limited evidence of a negative effect on savings, but this effect was not robust (i.e. did not persist with changes in data sets or specification).[14] This effect was relatively small in any case and the authors stress the many limitations of their results. In particular, their analysis cannot distinguish between the reduction of estates due to savings responses and those due to tax avoidance techniques.

Given the paucity of empirical evidence on the issue, the evidence on savings responses in general, and the theory outlined above, it appears difficult to argue for repeal of the estate tax to increase private saving. Even if the responsiveness to the estate and gift tax is as large as the largest empirical estimates of interest elasticities, the effect on savings and output would be negligible and more than offset by public dissaving.[15] Indeed, if the only objective were increased savings, it would probably be more effective to simply keep the estate and gift tax and use the proceeds to reduce the national debt.

Effect on Farms and Closely Held Businesses

Much attention been focused on the effect of the estate and gift tax on family farms and businesses and there is a perception that the estate tax is a significant burden on these businesses. Typically, family farm and business owners hold significant wealth in business and farm assets as well as other assets such as stocks, bonds and cash. Because many business owners are relatively well off and the estate and gift tax is a progressive tax, the probability of a farm or small business owner encountering tax liability is greater than for other decedents.[16]

Opponents of the estate and gift tax suggest that a family business or farm may in fact have to sell assets, often at a discounted price, to pay the tax. In his 1997 testimony, Bruce Bartlett from the National Center for Policy Analysis, stated that

> ...according to a survey, 51% of businesses would have difficulty surviving in the event of principal owner's death and 14% said it would be impossible for them to survive. Only 10% said the estate tax would have no effect; 30% said they would have to sell the family business, and 41% would have to borrow against equity.[17]

Are the data from this survey representative of the country as a whole? And, what are the policy issues associated with this effect? In response to the above testimony, there are two questions to explore. One, is repeal of the estate and gift tax efficiently targeted to relieve farms and small businesses? And two, of the farmer and small business decedents, how many actually encounter estate tax liability?

Target Efficiency

Congress has incorporated into tax law provisions, outlined earlier, that address and reduce the negative consequences of the estate tax on farms and small businesses. These laws are targeted to benefit only farm and small business heirs. In contrast, proposals to repeal the estate and gift tax entirely are poorly targeted to farms and small businesses.

Of the 18,430 taxable returns filed in 2005, 979 (5.3%) included farm assets. Additionally, no more than 8,273 (44.9%) returns included "business assets" in the estate.[18] (Note that some returns are double counted). Together, farm and business owners, by our definition, represent approximately 50.2% of all taxable estate tax returns.[19]

However, this estimate is dramatically overstated, even aside from the likelihood of double counting. The estimate for farms assumes any estate with a farm asset is a farm return thus including part-time farmers or those who may own farm land not directly farmed. The estimate for business assets may include many returns that include small interests (particularly for corporate stock and partnerships). Treasury data for 1998 indicated that farm estates where farm assets accounted for at least half of the gross estate accounted for 1.4% of taxable estates, while returns with closely held stock, non-corporate business or partnership assets equal to half of the gross estate accounted for 1.6%. The same data indicated that farm real estate and other farm assets in these returns accounted for 0.6% of the gross value of estates. Similarly estates with half of the assets representing business assets accounted for 4.1% of estates' gross values. Thus, it is clear that if the main motive for repealing the estate tax or reducing rates across-the-board were to assist farms and small businesses, most of the revenue loss would accrue to those outside the target group.

How Many Farm and Small Business Decedents Pay the Tax?

The more difficult question to answer is how many decedent farmers and small family business owners pay the tax. The first step in answering this question is to estimate the number of farmers and business owners (or those with farm and business assets) who die in any given year. We chose 2003 as our base year.

About 2.3 million people 25 and over died in the United States in 2004.[20] Some portion were farm and business owners. To estimate the number of those who died that were farm or business owners, we assume that the distribution of income tax filers roughly approximates the distribution of deaths in any given year. Or, the portion of farm individual income tax returns to total income tax returns in 2004 approximates the number farm deaths to total deaths. The same logic is used to approximate the number of business owner deaths. (Note that farmers tend to be older than other occupational groups and have somewhat higher death rates, which may slightly overstate our estimates of the share of farmer estates with tax).

In 2004, there were 132.2 million individual income tax returns filed; about 2 million were classified as farm returns and 20.3 million included business income or loss. These returns represent 1.5% and 15.3% of all returns respectively. If the profile of individual income tax return filers is similar to the profile of decedents, this implies approximately 35,267 farmers and 356,245 business owners died in 2004.[21]

Recall that the estate tax return data include 979 taxable returns with farm assets and 8,273 taxable returns we classify as business returns. Dividing these two numbers by the estimated number of deaths for each vocation yields an taxable estate tax return rate of 2.8% for farm owner decedents and 2.3% for business owner decedents. Thus, one can conclude that most farmers and business owners are unlikely to encounter estate tax liability.

Other Issues

Liquidity constraints or the inability of farms and small business to meet their tax liability with cash, may not be widespread. A National Bureau of Economic Research (NBER) paper using 1992 data estimated that 41% of business owners could pay estate and gift taxes solely out of narrowly defined liquid assets (insurance proceeds, cash and bank accounts); if stocks (equities) were included in a business's liquid assets, an estimated 54% could cover their estate and gift tax liability; if bonds are included an estimated 58% could cover their tax.[22]

These estimates suggest that only 3 to 4% of family farms and businesses would potentially be at risk even without accounting for the special exemptions; the special exemption suggests a much smaller number would be at risk.[23] If one included other non-business assets that are either not included in these estimates through lack of data (such as pensions) or nonfinancial assets (such as real estate) the estimate would be even higher. For many businesses a partial sale of assets (e.g. a portion of farm land) might be made or business assets could be used as security for loans to pay the tax. Finally, some estates may wish to liquidate the business because no heir wishes to continue it. Given these studies and analysis, it appears that only a tiny fraction, almost certainly no more than a percent or so, of heirs of business owners and farmers would be at risk of being forced to liquidate the family business to pay estate and gift taxes.

Effects of the Marital Deduction

One of the most important deductions from the estate tax is the unlimited marital deduction, which accounted for 30.1% of the gross value of all estates, and over 35% for larger estates (see table 3; larger estates may be more likely to reflect the death of the first spouse). An individual can leave his or her entire estate to a surviving spouse without paying any tax and getting step-up in basis (which permits no tax on accrued gains). The arguments for an unlimited marital deduction are obvious: since spouses tend to be relatively close in age, taxing wealth transferred between spouses amounts to a "double tax" in a generation and also discourages the adequate provision for the surviving spouse (although this latter objective could be met with a large, but not necessarily unlimited, marital deduction). (There is, however, a partial credit for prior transfers within a decade which could mitigate this double taxation within a generation.) Moreover, without an exclusion for assets transferred to the spouse, a substantial amount of planning early in the married couple's life (e.g. allowing for joint ownership of assets) could make a substantial difference in the estate tax liability.

Nevertheless, the unlimited marital deduction causes a certain amount of distortion. If a spouse leaves all assets to the surviving spouse, he or she forgoes the unified credit, equivalent to an exclusion that is currently $2.0 million and will eventually reach $3.5 million in 2009. In addition, because the estate tax is graduated, leaving all assets to a spouse can cause the couple to lose the advantage of going through the lower rate brackets twice. A very wealthy donor would leave enough to children (or to the ultimate beneficiary after the second spouse's death) to cover the exemption and to go through all of the rate brackets; then when the second spouse dies, another exemption and another "walk through the rate brackets" will be available. Donors can try to avoid the loss of these benefits and still provide for the surviving spouse by setting up trusts to allow lifetime income to the spouse and perhaps provide for invasion of the corpus for emergencies. These methods, of course, require pre-planning and may not be perfect substitutes for simply leaving assets to the surviving spouse, who would not have complete control.

Under other circumstances, the unlimited marital deduction can cause a decedent to leave more wealth to his or her spouse than would otherwise be preferable. For example, a decedent with children from a previous marriage might like to leave more assets to the children but the unlimited marital deduction may make it more attractive to leave assets to a spouse. One way of dealing with this problem is to leave a lifetime interest to the spouse and direct the disposal of the corpus of the trust to children. Indeed, the tax law facilitates this approach by allowing a trust called a Qualified Terminable Interest Property (QTIP) trust. Nevertheless, this approach also requires planning and is not a perfect substitute for directly leaving assets to children (particularly if the spouse has a long prospective life).

The point is that these provisions, whether deemed desirable or undesirable, distort the choices of a decedent and cause more resources to be devoted to estate planning than would otherwise be the case.

A Backstop for the Income Tax

Capital Gains

One reason frequently cited by tax analysts for retaining an estate tax is that the tax acts as a back-up for a source of leakage in the individual income tax — the failure to tax capital gains passed on at death. Normally, a capital gains tax applies on the difference between the sales price of an asset and the cost of acquiring it (this cost is referred to as the basis). Under current law, accumulated capital gains on an asset held until death will never be subject to the capital gains tax because the heirs will treat the market value at time of death (rather than original cost) as their basis. Assuming market values are estimated correctly, if heirs immediately sold these assets, no tax would be due. This treatment is referred to as "step-up in basis." It is estimated that 36% or more of gains escape taxes through step-up.[24]

The estate and gift tax is not a carefully designed back-up for the capital gains tax. It allows no deduction for original cost basis, it has large exemptions which may exclude much of capital gains from the tax in any case (including the unlimited marital deduction), and the tax rates vary from those that would be imposed on capital gains if realized. In particular, estate tax rates can be much larger than those imposed on capital gains (the current capital gains tax is capped at 20%, while the maximum marginal estate tax rate is 45%).

If the capital gains tax were the primary reason for retaining an estate and gift tax, then the tax could be restructured to impose capital gains taxes on a constructive realization basis. Alternatively, one could adopt a carry-over of basis, so that the basis remained the original cost, although this proposal could still allow an indefinite deferral of gain.

Owner-Occupied Housing, Life Insurance, and other Assets

Owner occupied housing and life insurance also escape income taxes on capital gains accrued through inside build-up (for the most part). Owner-occupied housing also escapes income tax on implicit rental income. There are practical economic and administrative reasons for some of these tax rules. It is administratively difficult to tax implicit rental income and taxing capital gains could potentially impede labor mobility. There are other assets as well that escape the income tax (such as tax exempt bonds). The estate and gift tax could also be seen as a backstop for these lapses in the individual income tax.

Effects of the Charitable Deduction

One group that benefits from the presence of an estate and gift tax is the nonprofit sector, since charitable contributions can be given or bequeathed without paying tax. As shown in table 3, estates that filed returns in 2005 donated 11.0% of total assets to charities; estates that occupy the highest wealth class ($20 million or greater) donated 24.3% of total assets to charity. Although one recent study found that charitable bequests are very responsive to the estate tax, and indeed that the charitable deduction is "target efficient" in the sense that it induces more charitable contributions than it loses in revenue, other studies have found a variety of responses, both small and large.[25] One problem with these types of studies is the difficulty in separating wealth and price effects.

An individual would have even greater tax benefits if charitable contributions were made during the lifetime, since they are deductible for purposes of the income tax, thereby reducing not only income tax but also, because the eventual estate is reduced, the estate tax as well. On the other hand, under the income tax charitable gifts are limited to 50% of income (30% for private foundations) and there are also restrictions on the ability to deduct appreciated property at full value. Despite this effect, a significant amount of charitable giving occurs through bequests and one study estimated that total charitable giving through bequests would fall by 12% if the estate tax were eliminated.[26] This reduction is, however, less than 1% of total charitable contributions.[27]

Charitable deductions play a role in some estate planning techniques described in the next section. In addition, some charitable deductions allow considerable retention of control by the heirs, as in the case of private foundations. Unlike the case of the income tax, there are no special restrictions on bequests to private foundations. (Under the income tax system, deductibility as a percent of income is more limited for gifts to foundations; there are also more limitations on gifts of appreciated property to foundations).

Efficiency Effects, Distortions, and Administrative Costs

A number of tax planning and tax avoidance techniques take advantage of the annual gift exclusion, the charitable deduction, the unlimited marital deduction, and issues of valuation. Because choices made with respect to these techniques can affect total tax liability, these planning techniques complicate compliance on the part of the taxpayer and administration on the part of the IRS. They may also induce individuals to arrange their affairs in ways that would not otherwise be desirable, resulting in distortions of economic behavior.

The most straightforward method of reducing estate and gift taxes is to transfer assets as gifts during the lifetime (inter-vivos transfers) rather than bequests. Gifts are generally subject to lower taxes for two reasons. First, assets can be transferred without affecting the unified credit because of the $12,000 annual exclusion. The exclusion was designed to permit gifts (such as wedding and Christmas presents) without involving the complication of the gift tax. This annual exclusion can, however, allow very large lifetime gifts. For example, a couple with two children, who are both married, could make $96,000 of tax-free gifts per year (each spouse could give $12,000 to each child and the child's spouse). Over 10 years, $960,000 could be transferred tax free (and without reducing the lifetime credit). Moreover, the estate is further reduced by the appreciation on these assets.

The effective gift tax is also lower than the estate tax because it is imposed on a tax-exclusive basis rather than a tax-inclusive basis. For example, if the tax rate is 45%, a taxable gift (beyond the $12,000 limit) of $100,000 can be given with a $45,000 gift tax, for an out-of-pocket total cost to the decedent of $145,000. However, if the transfer were made at death, the estate tax on the total outlay of $145,000 would be 45% of the total, or $65,250. Despite these significant advantages, especially from the annual exclusion, relatively little inter-vivos giving occurs.[28] There are a number of possible reasons for this failure to take advantage of the gift exclusion, and one is that the donee does not wish to relinquish economic control or perhaps provide assets to children before they are deemed to have sufficient maturity to handle them. There are certain trust and other devices that have been developed to allow some control to be maintained while utilizing the annual gift tax exclusion.[29] The annual gift exclusion can also be used to shift the ownership of insurance policies away from the person whose life is insured and out of the gross estate.

One particular method that allows a potentially large amount of estate tax avoidance is a Crummey trust. Normally, gifts placed in a trust are not eligible for the $12,000 exclusion, unless the trust allows a present interest by the beneficiary. The courts have held that contributions to a trust that allows the beneficiary withdrawal rights, even if the individual is a minor, and even if withdrawal rights are available for only a brief period (e.g. 15 or 30 days), can be treated as gifts eligible for the annual exclusion. This rule has been used to remove insurance assets from an estate (by placing them in a trust and using the annual $12,000 gift exclusion to pay the premiums without incurring tax). Under the Crummey trust, a large number of individuals (who may be children or other relatives of the primary beneficiaries) can be given the right (a right not usually exercised) to withdraw up to $12,000 over the limited time period. (Under lapse of power rules, however, this amount is sometimes limited to $6,000). All of these individuals are not necessarily primary beneficiaries of the trust but they expand the gift exclusion aggregate. In one case, a Crummey trust with 35 donees was reported.[30]

There is, however, one disadvantage of inter vivos gifts: these gifts do not benefit from the step-up in basis at death that allows capital gains to go unrecognized, so that very wealthy families with assets with large unrealized gains might prefer bequests (at least after the annual exclusion is used up).[31]

Individuals can also avoid taxes by skipping generations; although there is a generation skipping tax, there are large exemptions from the tax ($2.0 million per decedent in 2007, the same as the general estate tax exemption amount). Generation skipping may be accomplished through a direct skip (a decedent leaves assets to grandchildren rather than children) or an indirect skip (assets are left in a trust with income rights to children, and the corpus passing to the grandchildren on the children's death). The generation skipping tax rate is 45%. Relatively little revenue has been collected from the generation skipping tax because the tax has been successful in eliminating generation skipping transfers that are above the limit.[32]

Charitable deductions can also be used to avoid estate and gift taxes (and income taxes as well). For example, if a charity can be given rights to an asset during a fixed period (through a fixed annuity, or a fixed percentage of the asset's value), with the remainder going to the donor's children or other heirs, estate taxes can be avoided if the period of the trust is overstated (by being based on a particular individual life that is likely to be shorter than the actuarial life). Although restrictions have now been applied to limit reference persons to related parties, in the past so-called "vulture trusts" that recruited a completely unrelated person with a diminished life expectancy were used to avoid tax.[33]

Assets can also be transferred to charity while maintaining control through private foundations. Private foundations allow an individual or his or her heirs to direct the disposition of funds in the foundations for charitable purposes and continue to exercise power and control over the assets.

As noted earlier, some estate planning techniques are used to provide maximum benefits of the marital deduction plus the exclusion and lower rates; these approaches can also involve the use of trusts, such as the Qualified Terminable Interest Property (QTIP). These plans may permit the invasion of the corpus for emergency reasons.

Finally, a significant way of reducing estate taxes is to reduce the valuation of assets. A lower valuation can be achieved by transferring assets into a family partnership with many interests so that one party is not technically able to sell at a "market price" without agreement from the other owners to sell, a circumstance that the courts have seen as lowering the value of even obviously marketable assets, such as publicly traded stocks (the minority interest discount). Undervaluation can also be argued through the claim that a sale of a large block of stock (a "fire sale") would reduce asset value or, with a family-owned business, that the death of the owner (or a "key man") lowers the value substantially. A fractional interest in a property (such as real estate) may also qualify for a discount. Discounts may also be allowed for special use property whose market value may be higher than the value of the property in its current use.

Estate planning techniques complicate the tax law, increase the resources in the economy devoted to planning and also increase the administrative burden on the IRS especially when such cases go to court. Some claims have been made that the administrative costs and costs to taxpayers comprise a large part of the revenues. However, a recent study set the costs of complying with the estate tax at 6 to 9% of revenues. Moreover, an interesting argument was also made in that study that the inducement to settle affairs provided by the existence of an estate tax may be beneficial as it encourages individuals to get their affairs in order and avoid costly and difficult disputes among heirs.[34]

Of course, the administrative and compliance costs are, themselves, in part a consequence of the design of the tax. If the estate tax were revised to mitigate some of the need for tax planning, the administrative and compliance costs might be lower.

The high tax rates for some estates and the lack of third-party reporting mechanisms suggest that compliance may be a problem, although a large fraction of returns with large estates are audited. Estimates of the estate "tax gap," or the fraction of revenues that are not collected, have varied considerably; a recent estimate suggests about 13% of estates and gift taxes are not collected, although the authors suggest that this measure is very difficult to estimate.[35]

Repeal of Federal Credit for State Estate and Inheritance Taxes

In theory, the federal credit for state death taxes eliminated the incentive for states to "race to the bottom" of state estate tax rates and burden. Lower state liability simply increased federal liability by an equal amount. In short, the state credit was simply a federal transfer to states contingent upon the state's maintenance of an estate tax. The credit also reduced the federal tax burden of the estate and gift tax. The highest effective credit rate was of 16% of the gross estate value which reduced the highest federal rate of 55% to 39% (before EGTRRA).

In 2005, the "credit for state death taxes" was eliminated and replaced with a deduction for those taxes. Many states have relied on the federal credit for their estate tax and will need to modify their tax laws to continue collecting their estate and inheritance taxes. Under current state laws, "... there will be 29 states that have no state death tax in 2005."[36]

POLICY OPTIONS

Repealing the Estate and Gift Tax

One option is to eliminate the estate tax. This approach has been taken in the 2001 tax cut bill, the Economic Growth and Tax Relief Reconciliation Act of 2001 (EGTRRA, P.L. 107-16). However, the legislation sunsets after 2010, reverting to what the law would have been in 2011 if not for EGTRRA. During the phase-out period, the estate tax will still generate revenue, thus understating the full fiscal impact (revenue loss) of complete repeal. Immediate repeal of the estate and gift tax would cost up to $662 billion, whereas the estimated 10-year revenue cost of the temporary repeal of the estate tax under EGTRRA was $138 billion.

The Bush Administration's FY2007 budget proposal contained a proposal to permanently repeal the estate and gift tax beginning in 2010 by repealing the EGTRRA sunset. The revenue loss from that proposal would be $369.3 billion over the 2006-2016 budget window. Most of the revenue loss accrues in the out years of the proposal; from 2011 to 2016, the proposal would cost $334.1 billion in lost federal revenue.

Increasing the Credit, Converting to an Exemption, and/or Changing Rates

The two compromise proposals that were described earlier in this report would both increase the exemption (eliminating the credit structure) amount and lower the rate. Recall that the Kyl proposal from the 109th would set the exemption at $5 million for each spouse ($10 million per couple) and lower the tax rate to 15%. The Baucus proposal would set the exemption at $3.5 million and include a progressive rate schedule beginning at 15% and rising with the size of the estate to 35%. The relatively low rate of 15% is responsible for most the revenue loss under the Kyl proposal. No potential revenue loss estimates are available for the Baucus proposal, although it would be less expensive than the Kyl proposal given the graduated rate structure.

Under H.R. 5970, as passed in the 109th Congress, the tax rate on taxable estates up to $25 million would be equal to the tax rate on capital gains (currently 15% but scheduled to revert to 20% in 2011). The tax rate on estates valued over $25 million would drop to 30% by 2015. The JCT estimated that the estate tax provisions of H.R. 5970 would cost $268 billion over FYs 2007-2016, or about 69% of total repeal.[37]

For the same revenue cost, increasing the exemption would favor individuals with small estates and rate reductions would favor large estates. Indexing exemptions for inflation would preserve the value of the exemption against erosion by price inflation.

Taxing the Capital Gains of Heirs vs. the Estate Tax

In 2010, the estate tax will be replaced by a tax on the capital gains of heirs when an inherited asset is sold. The new law includes an exemption from carryover basis for capital gains of $1.3 million (and an additional $3 million for a surviving spouse). There are efficiency losses to taxing capital gains of heirs on inherited assets because such taxation would increase the lock-in effect. The lock-in effect occurs when potential taxpayers hold onto their assets because the anticipated tax on the gain. If the asset value grows from generation to generation, the lock-in effect becomes stronger and stronger. Some analysts have suggested that the result of the lock-in effect will be familial asset hoarding. Legislation that repeals the EGTRRA sunset would retain this treatment of capital gains.

Both taxation of gains at death and carry-over basis may be complicated by lack of information by the executor on the basis of assets (some of which may have been originally inherited by the decedent). Indeed, a proposal in the 1970s to provide carry-over basis was never put into place because of protests, some associated with the problem of determining the basis. This problem may be less serious for EGTRRA because of the carry-over of basis exemption.

Allowing constructive realization or carryover basis may further complicate estate planning if an exemption were allowed, because it would be advantageous to pick those assets with the largest amounts of appreciation for the exclusion or carryover basis. In addition, since the tax arising from carryover basis would depend on the heir's income tax rates, revenues could be saved by allocating appreciated assets to heirs with the lowest expected tax rates.

Changing from the current step-up basis for inherited assets to a carryover basis, as enacted by EGTRRA, will also affect the life insurance choices of taxpayers. Generally, EGTRRA will likely encourage taxpayers to invest more in life insurance than other investments. Under current tax law, the appreciation of assets held in life insurance policies is not subject to capital gains taxes. Also, the payout from these policies, usually in cash to heirs, is not subject to income taxes and effectively receives a stepped up basis. The switch to carryover basis at death would then favor life insurance policies over other assets that are subject to capital gains taxes at death (assuming the heir liquidates the assets). The anticipated change to more assets held in life insurance policies will likely reduce the revenue generated by capital gains taxes.

Concerns of Farms and Family Businesses

If the estate tax is not repealed, farms and family businesses may be targeted for further relief by increases in the special exemptions for farm and business assets.[38] Because of the asset distribution of estates, almost all decedents with significant farm or business assets would likely not pay an estate tax. While this approach would be effective at targeting family farms and businesses for relief, it would exacerbate a general concern with the estate and gift tax — the unfairness of a differential treatment of owners of these business and farm assets compared to those with other forms of assets. Why should a wealthy individual whose assets are in a closely held corporation escape estate and gift tax on his or her assets, while an individual who holds shares in a publicly traded corporation pay a tax?

Larger exemptions also encourage wealthy decedents to convert other property into business or farm property to take advantage of the special exemptions. An incentive already exists to shift property into this exempt form and it would have been exacerbated by an expansion of the exemption.

Reform Proposals and other Structural Changes

Many pre-EGTRRA proposals were intended to modify rather than completely repeal the estate tax. Some of these proposals may be revisited as the subset provision in EGTRRA nears. The proposed revisions would have focused on eliminating estate tax avoidance schemes or at fixing current inconsistencies in the estate tax law. The issues presented here are still relevant for the near term for two reasons. One, during the phase-out period, the estate and gift is still part of the tax code. Two, the gift tax is retained even after eventual repeal of the estate tax.

Phase out of Unified Credit for Largest Estates
This provision would allow for a phase out of the unified credit as well as the lower rates, by extending the bubble. This provision would cause all assets in large estates to be taxed at the top rate of 47%.

Impose Consistent Valuation Rules
Analysts have proposed that valuation of assets be the same for income tax purposes as for estate tax purposes. Basically, it is advantageous to *overvalue* assets for purposes of the income tax, so as to minimize any future capital gains tax liability. Conversely, its advantageous to *undervalue* assets to minimize estate tax liability. In addition to requiring consistency in valuation, some have proposed that donors report the basis of assets given as gifts. Currently, assets transferred by gift and then sold do not benefit from step-up in basis even though the donor is not required to report the basis to the donee.

Modify the Rules for the Allocation of Basis
Under current law, a transaction that is part gift and part sale assigns a basis to the asset for the donee that is the larger of the fair market value or the amount actually paid. The donor pays a tax on the difference between amount paid and his basis and may frequently recognize no gain. This proposal would allocate basis proportionally to the gift and sale portions.[39]

Eliminate Stepped up Basis on Survivor's Share of Community Property
Under present law, in common law states, half of property held jointly by a married couple is included in the first decedent's gross estate and that half is thus eligible for step-up in basis for purposes of future capital gains. In community property states, however, where all properties acquired during marriage are deemed community property, a step up in basis is available for all community property, not just the half that is allocated to the decedent spouse. The reason for this rule, which is quite old, was the presumption in the past that property in a common law state would have been held by the husband (who would have acquired it) and thus would all have been eligible for step-up, while only one half of property in community property states would have been deemed to be held by the husband and be eligible for step up. This older treatment, it is argued, was made obsolete by changes in 1981 that determined that only half of any jointly held property would be included in the estate regardless of how the property was acquired, and thus made the step up apply to only one half of this type of property. Thus, currently couples in community property states are being treated more favorably than those in common law states.

A reservation with this treatment is that property that could be allocated to one spouse in a common law state may not be able to escape the community property treatment in a community property states, and common law states may now be favored if these assets in common law states tend to be held by the first decedent. However, couples in community property states may be able to convert to separate property by agreement, and thereby take advantage of the same planning opportunities as those in common law states.

Modify QTIP Rules

Under present law, an individual may obtain a marital deduction for amounts left in trust to a spouse under a Qualified Terminable Interest Property (QTIP) trust, with one requirement being that the second spouse must then include the trust amounts in their own estate. In some cases the second estate has argued that there is a defect in the trust arrangement so that the trust amount is not included in the second spouse's estate (even though a deduction was allowed for it in the first spouse's estate). This provision would require inclusion in the second spouse's estate for any amount excluded in the first spouse's estate.

Eliminate Non-Business Valuation Discounts

This provision would require that marketable assets be valued at the fair market value (i.e., there would be no valuation discounts for holding assets in a family partnership or for "fire-sale" dispositions).

Eliminate the Exception for a Retained Interest in Personal Residences from Gift Tax Rules

Under current law, when a gift is made but the grantor retains an interest, that retained interest is valued at zero (making the size of the gift and the gift tax larger). In the case of a personal residence, however, the retained interest is valued based on actuarial tables. In general, retained interests are only allowed to be deducted from the fair market of the gift (reducing gift taxes) if they can be objectively valued (and hence are allowed for certain types of trusts, such as those that pay an annuity).

Disallow Annual Gift Taxes in a Crummey Trust

As noted earlier, the annual gift exclusion is not available for gifts placed in trust unless certain rules are met, but a Crummey trust, which allows some right of withdrawal, is eligible. This revision would allow gifts in trust to be deductible only if the only beneficiary is the individual, and if the trust does not terminate before the individual dies, the assets will be in the beneficiary's estate. These rules are similar to generation skipping taxes.

Reduce the Annual Gift Tax Exclusion

The annual gift tax exclusion allows significant amounts to be transferred free of tax and also plays a role in transferring insurance out of the estate (by using the annual gift tax exclusion to pay the premium). While some gift tax exclusion is probably desirable for simplification purposes, the $12,000 exclusion's role in estate tax avoidance could be reduced by reducing its size. An alternative change that would limit the use of the annual exclusion in tax avoidance approaches would be a single exclusion per donor (or some aggregate limit per donor), to prevent the multiplication of the excluded amount by gifts to several children and those children's spouses and the use of techniques such as the Crummey trust.

Allow Inheritance of Marital Deductions or Lower Rates

One of the complications of estate planning is maximizing the use of the exemption and lower rate brackets by a married couple. In this case, while it may be economically and personally desirable to pass the entire estate (or most of the estate) to the surviving spouse, minimizing taxes would require passing to others at least the exemption amount and perhaps more to take some advantage as well of the lower rate brackets. Such complex situations could be avoided by allowing the surviving spouse to inherit any unused deduction and lower rate brackets so that the couple's full deductions and lower rates could be utilized regardless of how much was left to the surviving spouse.

Allow Gift Tax Treatment Only on Final and Actual Transfer

Many of the tax avoidance techniques with charitable gifts involve over-valuation of a deductible interest. For example, a gift may be made of the remainder interest after an annuity has been provided to a charitable organization . The larger the value of the charitable annuity, the smaller the value of the gift (and the gift tax). One way to over-value an annuity is to allow the annuity to extend to a particular individual's lifetime, when that individual has a shorter life than the actuarial tables indicate. Similarly, a way to transfer income via the gift tax exclusion without the recipient having control over it is to place it in a Crummey trust. Even if the individual is able to exercise withdrawal rights, the expectation of not receiving future gifts if the assets are withdrawn in violation of the donor's wishes may mean that such rights will never be exercised. A rule that excludes all assets placed in trusts from consideration for the gift tax would eliminate this mechanism.

These types of valuation techniques could be addressed by only allowing the gift tax to be imposed at the time of the actual final transfer. In such a case, no actuarial valuation would be necessary and no trust mechanisms would be available.

Valuation of Assets

One option is to disallow discounts for property that has a market value (such as bonds and publicly traded stock) regardless of the form the asset is held in, as suggested by the administration tax proposals. Such a change would prevent the avoidance technique of placing assets into a family partnership or similar arrangement and then arguing that the property has lost market value because it would require agreement of the heirs to sell it. In addition, other limits on valuation discounts could be imposed. For example, blockage discounts based on "fire sale" arguments could be disallowed. Such a provision might allow for an adjustment if the property is immediately sold at such a lower price.

Include Life Insurance Proceeds in the Base

Some tax avoidance techniques are associated with shifting life insurance proceeds out of the estate by shifting to another owner.

Switch to an Inheritance Tax

Some authors have suggested that an inheritance tax should be substituted for the estate tax. Some states have inheritance taxes. An estate tax applies to the total assets left by the decedent. An inheritance tax would be applied separately to assets received by each of the heirs. If tax rates are progressive, smaller taxes would be applied the greater the number of beneficiaries of the assets. One reason for such a change would, therefore, be to encourage more dispersion of wealth among heirs, since taxes would be lower (assuming exemptions and graduated rates) if split among more recipients. In addition, under an inheritance tax the tax rate can be varied according to the status of the heir (son vs. cousin, for example). At the same time, one can see more avoidance complications arising from an inheritance tax.

CONCLUSION

The analysis in this report has suggested that some of the arguments used for and against maintaining the estate tax may be questioned or of lesser import than is popularly assumed. For example, there is little evidence that the estate tax has much effect on savings (and therefore on output); indeed, estate taxes could easily increase rather than reduce savings. Similarly, only a tiny fraction of farms and small businesses face the estate and gift tax and it has been estimated that the majority of those who do have sufficient non-business assets to pay the tax. Moreover, only a small potion of the estate tax is collected from these family owned farms and small businesses, so that dramatically reducing estate tax rates or eliminating the tax for the purpose of helping these family businesses is not very target efficient.

Although the estate tax does contribute to the progressivity of the tax system, this progressivity is undermined, to an undetermined degree, by certain estate tax avoidance techniques. Of course, one alternative is to broaden the estate tax base by restricting some of these estate planning techniques. At the same time progressivity could be achieved by other methods.

On the other hand, arguments that the estate tax is a back-up for the income escaping the capital gains tax, would not support the current high rates of the estate tax, which should be lowered to 20% or less to serve this purpose.

More intangible arguments, such as the argument that inheritances are windfalls that should be taxed at higher rates on the one hand, or that death is an undesirable time to levy a tax and that transferred assets have already been subject to taxes, are more difficult to assess but remain important issues in the determination of the desirability of estate and gift taxes.

APPENDIX: ESTATE AND GIFT TAX DATA

Table A1. The Filing Requirement and Unified Credit

Year of Death	Filing Requirement or Equivalent Exemption	Unified Credit
2004 and 2005	$1,500,000	$555,800
2006 through 2008	$2,000,000	$780,800
2009	$3,500,000	$1,525,800
2010	estate tax repealed	estate tax repealed
2011 and after	$1,000,000	$345,800

Table A2. Gross Estate Value of Taxable Returns Filed in 2005

Size of Gross Estate	All Returns	Taxable Returns	Gross Estate Value (in 000s)	Gross Taxable Estate Value (in 000s)	Percent Taxable Estate Tax Returns
All Returns	39,482	18,430	192,635,099	101,771,906	46.68%
$1.5 to $2.5 million	21,347	8,668	70,145,137	16,866,733	40.61%
$2.5 to $5.0 million	11,895	6,162	36,984,942	20,763,258	51.80%
$5.0 to $10.0 million	4,122	2,280	25,957,237	15,590,318	55.31%
$10.0 to $20 million	1,358	822	17,906,950	11,251,943	60.53%
over $20.0 million	760	498	41,640,833	37,299,654	65.53%

Source: Internal Revenue Service, Statistics of Income, *Estate Tax Returns Filed in 2005*, IRS, SOI unpublished data, November 2006.

Table A3. Allowable Deductions on 2005 Returns (Sorted by Total Value)

Deduction	Returns with Deduction		Value of Deductions (in millions)	
	Total	Taxable	Total	Taxable
Total deductions	$39,445	$18,396	$83,082,614	$28,508,260
Bequests to surviving spouse	$18,224	$1,708	$53,679,406	$8,915,127
Charitable deductions	$8,074	$4,344	$19,563,951	$13,532,457
Debts and mortgages	$29,108	$16,244	$6,311,729	$3,176,059
Executor's commissions	$13,691	$11,192	$1,081,099	$957,111
Other expenses and losses	$24,284	$15,822	$1,022,600	$893,681
Attorney's fees	$24,061	$15,964	$869,710	$685,026
Funeral expenses	$34,356	$17,599	$326,382	$161,877
State death taxes paid	$673	$400	$119,740	$104,710

Source: Internal Revenue Service, Statistics of Income, *Estate Tax Returns Filed in 2005*, IRS, SOI unpublished data, November 2006.

Table A4. 2007 Estate Tax Rate Schedule

Taxable Estate Value From	to	Current Statutory Rate (in Percent)
$0	$10,000	18
$10,001	$20,000	20
$20,001	$40,000	22
$40,001	$60,000	24
$60,001	$80,000	26
$80,001	$100,000	28
$100,001	$150,000	30
$150,001	$250,000	32
$250,001	$500,000	34
$500,001	$750,000	37
$750,001	$1,000,000	39
$1,000,001	$1,250,000	41
$1,250,001	$1,500,000	43
$1,500,001	and over	45

Table A5. Repealed Credit for State Death Taxes

Taxable Estate Value (less the $60,000 exemption)	to	Current Statutory Credit Rate (in Percent)
$0	$40,000	0
$40,001	$90,000	.8
$90,001	$140,000	1.6
$140,001	$240,000	2.4
$240,001	$440,000	3.2
$440,001	$640,000	4.0
$640,001	$840,000	4.8
$840,001	$1,040,000	5.6
$1,040,001	$1,540,000	6.4
$1,540,001	$2,040,000	7.2
$2,040,001	$2,540,000	8.0
$2,540,001	$3,040,000	8.8
$3,040,001	$3,540,000	9.6
$3,540,001	$4,040,000	10.4
$4,040,001	$5,040,000	11.2
$5,040,001	$6,040,000	12.0
$6,040,001	$7,040,000	12.8
$7,040,001	$8,040,000	13.6
$8,040,001	$9,040,000	14.4
$9,040,001	$10,040,000	15.2
$10,040,001	and over	16.0

Table A6. Wealth Distribution of Taxable Returns Filed in 2005

Size of Gross Estate	Taxable Returns	Gross Taxable Estate Value (millions)	Net Estate Tax (millions)	Percent of Taxable Estate Returns	Percent Federal Estate Tax Value
All Returns	$18,430	$101,771,906	$21,520,989	100.00%	100.00%
1.5 to 2.5 million	$8,668	$16,866,733	$1,550,048	47.03%	16.57%
2.5 to 5.0 million	$6,162	$20,763,258	$4,393,227	33.43%	20.40%
5.0 to 10.0 million	$2,280	$15,590,318	$4,477,023	12.37%	15.32%
10.0 to 20 million	$822	$11,251,943	$3,275,972	4.46%	11.06%
over 20.0 million	$498	$37,299,654	$7,824,719	2.70%	36.65%

Note. In 2005, the credit was repealed and replaced with a deduction.
Source: Internal Revenue Service, Statistics of Income, *Estate Tax Returns Filed in 2005*, IRS, SOI unpublished data, November 2006.

REFERENCES

[1] Joint Committee on Taxation, *Estimated Revenue Effects of H.R. 8, the Death Tax Permanency Act of 2005*, JCX-20-05, 109th Congress, April 13, 2005.

[2] Kurt Ritterpusch, "Baucus Proposal could Complicate Effort in senate to Find 60 Votes for Repeal Plan," *Daily Tax Report*, no. 105, June 1, 2006.

[3] Joel Friedman, "Estate Tax 'Compromise' with 15 Percent Rate is Little Different Than Permanent Repeal," *Center on Budget and Policy Priorities*, May 31, 2006.

[4] Possible consequences that have been discussed include concentrations of political power, inefficient investments by the very wealthy, and disincentives to work by heirs (often referred to as the Carnegie conjecture, reflecting a claim argued by Andrew Carnegie).

[5] For a history of the estate and gift tax as well as a detailed explanation of current law, see the following CRS reports by John R. Luckey: CRS Report 95-416, *Federal Estate, Gift, and Generation-Skipping Taxes: A Description of Current Law*, and CRS Report 95-444, *A History of Federal Estate, Gift, and Generation-Skipping Taxes*.

[6] 26 I.R.C. 2001(c)

[7] The federal credit for state death taxes paid was repealed beginning in 2005. See Table A4 in the appendix for the old credit for state death taxes paid schedule.

[8] We have dropped the modifier "adjusted" from taxable estate for the benefit of the reader. The taxable estate and the adjusted taxable estate are identical in the absence of taxable gifts.

[9] The data are from those estates that filed in 2005, thus many estates followed the 2004 rules which still included the credit for state death taxes.

[10] This is slightly greater than the tentative estate tax less credits because of rounding.

[11] Mortality data for 2004 is the latest year available.

[12] For more see Douglas Holtz-Eakin, David Joulfaian, and Harvey Rosen, "The Carnegie Conjecture: Some Empirical Evidence," *Quarterly Journal of Economics*, v. 108, May 1993, pp. 413-435.

[13] William G. Gale and Maria G. Perozek. Do Estate Taxes Reduce Savings? April 2000. Presented at a Conference on Estate and Gift Taxes sponsored by the Office of Tax Policy Research, University of Michigan, and the Brookings Institution, May 4-5, 2000.

[14] Wojciech Kopczuk and Joel Slemrod, The Impact of the Estate Tax on the Wealth Accumulation and Avoidance Behavior of Donors. April 17, 2000. Presented at a Conference on Estate and Gift Taxes sponsored by the Office of Tax Policy Research, University of Michigan, and the Brookings Institution, May 4-5, 2000.

[15] Interest elasticities have been estimated at no higher than 0.4; that is, a one percent increase in the rate of return would increase savings by 0.4%. Ignoring the effect on the deficit or assuming the revenue loss is made up by some other tax or spending program that has no effect on private savings, this amount is about 40% of the revenue cost, so that savings might initially increase by about $12 billion. Output would rise by this increase multiplied by the interest rate, or about $1 billion (or, 1/100 of 1% of output). In the long run, savings would accumulate, and national income might eventually increase by about one tenth of 1%. (This calculation is based on the following: the current revenue cost of $28 billion accounts for about 1.4% of capital income of

approximately 25% of Net National Product; at an elasticity of 0.4, a 1.4% increase in income would lead to a 0.56 % increase in the capital stock and multiplying by the capital share of income (0.25) would lead to an approximate 0.14 increase in the capital stock.)

[16] See CRS Report RL33070, *Estate Taxes and Family Businesses: Economic Issues*, by Jane Gravelle and Steven Maguire.

[17] Statement before the Subcommittee on Tax, Finance, and Exports, Committee on Small Business, June 12, 1997.

[18] A return is classified as a business return if at least one of the following assets is in the estate: closely held stock, limited partnerships, real estate partnership, other non-corporate business assets. Counting the same estate more than once is likely which significantly overstates the number of business estate tax returns.

[19] See CRS Report RS20593, *Asset Distribution of Taxable Estates: An Analysis*, by Steven Maguire.

[20] Mortality data for 2004 is the latest year available.

[21] The percentages are multiplied by the 2,374,781 deaths of those 25 years old and over. If the age were higher then the pool of decedents would be smaller and the percentage that paid estate taxes incrementally higher.

[22] Holtz-Eakin, Douglas, John W. Philips, and Harvey S. Rosen, "Estate Taxes, Life Insurance, and Small Business," *National Bureau of Economic Research*, no. 7360, September 1999, p. 12.

[23] Of course, if all heirs do not wish to continue ownership in the family business, these liquid assets might need to be used to buy them out; that, however, is a choice made by the heirs and not a forced sale.

[24] See Poterba, James M. and Scott Weisbenner, "The Distributional Burden of Taxing Estates and Unrealized capital Gains at the Time of Death," *National Bureau of Economic Research*, no. 7811, July 2000, p. 36.

[25] See David Joulfaian, Estate Taxes and Charitable Bequests by the Wealth, National Bureau of Economic Research Working Paper 7663, April 2000. This paper contains a review of the econometric literature on the charitable response. Note that, in general, a tax incentive induces more spending than it loses in revenue when the elasticity (the percentage change in spending divided by the percentage change in taxes) is greater than one.

[26] Ibid.

[27] Bruce Bartlett, Misplaced Fears for Generosity, *Washington Times*, June 26, 2000, p. A16.

[28] See James Poterba. "Estate and Gift Taxes and Incentives for *Inter Vivos* Giving in the United States," forthcoming *Journal of Public Economics*. Even at high income levels, Poterba found that only about 45% of households take advantage of lifetime giving. He also found that those with illiquid assets (such as family businesses) and those with large unrealized capital gains are less likely to make inter-vivos gifts.

[29] For a more complete discussion of this and other techniques, see Richard Schmalbeck, Avoiding Wealth Transfer Taxes, paper presented at the conference Rethinking Estate and Gift Taxation, May 4-5, 2000, Office of Tax Policy Research, University of Michigan, and the Brookings Institution and Charles Davenport and Jay Soled. Enlivening the Death-Tax Death-Talk. *Tax Notes*, July 26, 1999, pp. 591-629.

[30]   See Davenport and Soled, *op cit.*

[31]   See David Joulfaian, Choosing Between Gifts and Bequests: How Taxes Affect the Timing of Wealth Transfers. U.S. Department of Treasury Office of Tax Analysis Paper 86. May 2000.

[32]   For a description see CRS Report 95-416, *Federal Estate, Gift, and Generation-Skipping Taxes: A Description of Current Law*, by John R. Luckey.

[33]   See Schmalbeck, *op cit.* The reference individual cannot be terminal (have a life expectancy of less than a year), however.

[34]   See Davenport and Soled, *op cit.* This study also reviewed a variety of other studies of estate tax compliance costs.

[35]   See Martha Eller, Brian Erard and Chih-Chin Ho, *The Magnitude and Determinants of Federal Estate Tax Noncompliance*, Rethinking Estate and Gift Taxation, May 4-5, 2000, Office of Tax Policy Research, University of Michigan, and the Brookings Institution.

[36]   Harley Duncan, "State Responses to Estate Tax Changes Enacted as Part of the Economic Growth and Tax Relief Reconciliation Act of 2001 (EGTRRA)," *State Tax Notes*, Dec. 2, 2002, p. 615.

[37]   Joint Committee on Taxation, *Estimated Revenue Effects of H.R. 5970, The "Estate Tax and Extension of Tax Relief Act of 2006 ('ETETRA'),"* as introduced in the House of Representatives on July 28, 2006, JCX-34-06, 109th Congress, July 28, 2006.

[38]   For more on the estate tax and its effect on family businesses, see CRS Report RL33070, *Estate Taxes and Family Businesses: Economic Issues*, by Jane Gravelle and Steven Maguire.

[39]   For example, suppose an asset with a basis of $50,000 but a market value of $100,000 is sold to the donee for $50,000. The donor would realize no gain and the gift amount would be $50,000, with the donee having a basis of $50,000. Eventually, the gain would be taxed when the donee sold the property, but that tax would be delayed. However, if the asset were divided into a gift of $50,000 with a basis of $25,000 and a sale of $50,000 with a basis of $50,000, the donor would realize a gain of $25,000; the donee would now have a basis of $75,000. Half of the gain would be subject to tax immediately.

In: Agricultural Finance and Credit
Editor: J. M. Bishoff, pp. 31-46

ISBN: 978-1-60456-072-5
© 2008 Nova Science Publishers, Inc.

Chapter 2

ESTATE TAXES AND FAMILY BUSINESSES: ECONOMIC ISSUES[*]

Jane G. Gravelle[1] and Steven Maguire[2]

[1] Senior Specialist in Economic Policy Government and Finance Division
[2] Analyst in Public Finance Government and Finance Division

ABSTRACT

The 2001 tax revision began a phaseout of the estate tax, by increasing exemptions and lowering rates. The estate tax is scheduled to be repealed in 2010 and a provision to tax appreciation on inherited assets (in excess of a limit) will be substituted. The 2001 tax provisions sunset, however, so that absent a change making them permanent the estate tax will revert, in 2011, to prior, pre-2001, law. Proposals to make the repeal permanent, or to significantly increase the exemptions and lower the rate, are under consideration.

Currently, discussions of the estate tax are focusing particular attention on the effects on family businesses, including farms, and perception that the estate tax unfairly burdens family businesses because much of the estate value is held in illiquid assets (e.g., land, buildings, and equipment). The estate tax may even force the liquidation of family businesses. A special family business deduction, the Qualified Family Owned Business Interest Exemption (QFOBI) was enacted in 1997. Presently, because of higher exemptions allowed and a previous cap on the combined regular and small business exemption, this provision is no longer relevant. If, however, the estate tax repeal sunsets, QFOBI will again be germane. In the 109th Congress, H.R. 8, which would make the estate tax repeal permanent, was passed by the House, but not by the Senate. There were also proposals to allow an expanded business exemption (H.R. 1612 and S. 928) as well as proposals to allow a higher exemption (H.R. 1577 and H.R. 1574) or both a higher exemption and lower rate (H.R. 1560, H.R. 1568, H.R. 1614, and H.R. 5638). H.R. 5970 — a proposal for a credit eventually equivalent to a $5 million exemption ($10 million for a married couple) with tax rates initially set at the capital gains tax rate (currently 15%, and scheduled to rise to 20%) for estates up to $25 million, and at twice the gains rate for those over $25 million — was passed by the House on July 29, 2006.

---

[*] Excerpted from CRS Report RL33070, dated January 26, 2007.

Evidence suggests, however, that only a small fraction of estates with small or family business interests have paid the estate tax (about 3.5% for businesses in general, and 5% for farmers, compared to 2% for all estates). Recent estimates suggest that only a tiny fraction of family-owned businesses (less than ½ of 1%) are subject to the estate tax but do not have readily available resources to pay the tax. Thus, while the estate tax may be a burden on those families, the problem is confined to a small group.

If the estate tax is repealed, QFOBI will allow an exemption for some or all of business assets in about a third to a half of estates with more than half their assets in these businesses, but the value of the exemption will be reduced because the general exemption has increased. If the estate tax repeal is made permanent, liquidity will cease being a problem, although family businesses may be more likely than other estates to be affected by the capital gains provisions. Exposure to the estate tax, if it is reinstated, would be significantly decreased by increases in either the family business or general exemptions. The report also discusses an uncapped exemption and an uncapped exemption targeted at liquidity issues. This report will be updated as legislative events warrant.

The Economic Growth and Tax Relief Reconciliation Act of 2001 (EGTRRA) phased out the estate tax by gradually increasing the exemption and lowering the rate; the tax will be eliminated in 2010. A provision providing for carryover basis for assets transferred at death will replace the current step-up in basis. This latter provision would require heirs, when selling inherited assets, to pay tax on the gain that existed at the time of death, in addition to any appreciation since transfer. (A $1.3 million exemption would be allowed, with an additional $3 million for a surviving spouse.) The repeal of the estate tax and the provision for carryover basis is sunsetted and, absent legislative change, will revert to pre-2001 law in 2011, with an exemption of $1 million and a top statutory tax rate of 55%. The law will also revert to the prior rule of a stepped-up basis for assets, where no capital gains tax would be paid on appreciation of assets existing at the time of death.

Currently, discussions about the estate tax are focusing particular attention on the effects on family businesses, including farms.[1] Many policy makers and observers have maintained that the estate tax unfairly burdens family businesses because much of the estate value is held in illiquid assets (e.g., land, buildings, and equipment).[2] The estate tax, it is suggested, forces these businesses to liquidate vital assets to pay the tax. Critics of the estate tax posit that, in some cases, liquidating the business completely may be the only option.

Prior law contained a special deduction for family-owned businesses (the qualified family-owned business interest deduction, or QFOBI) to address this issue; this deduction was capped so that the total of the normal exemption and the family business exemption could not exceed $1.3 million. The deduction was also contingent on meeting a number of qualifying rules. QFOBI is currently irrelevant since the exemption is $1.5 million (and was repealed by EGTRRA for 2004 and beyond), but will play a role again if the EGTRRA provisions sunset. The estate tax (both under current and permanent rules) contains some other provisions that may make the payment of the tax easier for family-owned businesses. Qualifying estates can pay the tax in installments over a maximum of 10 years after a five-year deferral, and can value their land as currently used rather than at fair market value.[3] Family-owned business may also be more likely to benefit from minority and marketability discounts, which allow a lower valuation if the property is held by several heirs with a minority interest or is otherwise difficult to sell.[4]

Proposals have been made to make the 2010 provision permanent, and legislation to that effect in the 109th Congress, H.R. 8, passed the House. Making the change permanent would involve a significant revenue loss, in excess of $50 billion in 2012.[5] There were also proposals to retain the estate tax but allow an expanded business exemption (H.R. 1612 and S. 928), a higher exemption (H.R. 1577 and H.R. 1574), or both a higher exemption and lower rate (H.R. 1560, H.R. 1568, and H.R. 1614).[6] Recently proposals have been made to retain the estate tax; they include a proposal suggested by Senator Kyl to provide a $5 million exemption per spouse and a 15% rate (equal to the current capital gains tax rate). Senator Baucus has a proposal to provide a larger exemption and lower, but graduated, rates as a permanent provision.

In 2006, the House adopted H.R. 5638, which allowed a $5 million exemption per spouse and sets the rate at the capital gains tax rate (currently 15% but scheduled to revert to 20% in 2011) for estates not over $25 million and twice that rate for the remainder; this legislation was included in H.R. 5970, passed by the House, and now being considered by the Senate.

This report discusses the general issue of family-owned businesses, and then discusses the consequences of several options including making no revisions (and hence returning to pre-2001 law in 2011); making the repeal of the estate tax permanent; and, modifying exemptions while retaining an estate tax. This latter discussion considers expanding or altering the existing general or business exemptions, providing an exemption for all business assets (no dollar ceiling or other restrictions), or modifying the QFOBI-type exemption to eliminate a cliff effect that currently exists because only estates with half of the assets in a business were eligible for the special deduction.

FAMILY BUSINESS ASSETS IN THE ESTATE TAX BASE

Although much attention has been devoted to the effect of the estate tax on family farms and businesses, and in particular the forced liquidation of family businesses and business assets, very few family businesses pay the estate tax and very little of the estate tax is collected from family businesses (and none from truly small businesses due to the exemption levels). Of those estates with family business assets, most would appear to be able to pay the tax, as discussed below.

Farm and business assets appear to account for around 11-12% of taxable estate assets, and perhaps a little more because these data are from 2003 when some QFOBI still existed.[7]

Some data reported in another CRS report[8] are instructive regarding the ability of family-owned businesses to pay the tax. Of taxable estates in 2003, 6.4% reported farm assets and 36.0% reported business assets. However, while around 40.0% of taxable estate tax returns report some business or farm assets (there is some overlap between the farm and business numbers), only 1.4 % of estates were those where farm assets were at least half the estate and only 1.6% of estates had business assets that accounted for half the estate. Farm assets in these returns accounted for 0.6% of estate value and business assets 4.1%. Since estates with less than half their assets in farm or business assets would have other sources (from the estate itself) to pay the tax, only this small fraction of estates would presumably have to deal with possible liquidation, and these estates were targeted by QFOBI. Table 1 and table 2 below report the percentage of estates with business assets by estate size and the value of the business assets as a percentage of the estate by estate size for returns filed in 2003, respectively.

This same report also estimated that only 3.3% of business owners face the estate tax, and about 5% of farmers do. This share is slightly larger than the share of all decedents, where less than 2% pay the tax, reflecting the higher average wealth of the farm and business owners.

If one considers the ability to pay the tax out of all non-business assets, then if half the estate is held in other assets, given a tax rate of 55%, and since assets are excluded from the estate through general exemptions as well, the non-business assets would generally be adequate to pay the tax (since the tax as a share of the estate will usually be below 50% of the value). According to the data reported above, about 20% of returns with farm assets (1.4/6.4) have more than 50% of assets in farm assets, suggesting that less than 1% of decedents with farm assets would not have enough non-business resources to pay the tax. For business owners, only about 4% involve returns where business assets are more than one half the estate. Thus, by this calculation, less than 1/10 of 1% of business owners would not have enough nonbusiness resources to pay the tax.

Table 1. Percentage of Estates with Business Assets by Type of Asset and Estate Size in 2003

Size of Gross Estate	Number	Real Estate Partnerships	Closely Held Stock	Farm Assets	Limited Partnerships	Other Non-corp. Bus. Assets
All Returns	66,043	3.8%	14.4%	6.4%	12.0%	9.6%
1 to 2.5 million	49,748	2.5%	11.0%	6.4%	9.2%	7.3%
2.5 to 5.0 million	10,549	5.3%	20.1%	6.4%	16.2%	12.8%
5.0 to 10.0 million	3,732	9.5%	28.1%	6.6%	24.2%	19.0%
10.0 to 20 million	1,293	16.8%	37.5%	7.5%	34.3%	27.8%
over 20.0 million	721	20.8%	50.6%	10.8%	44.4%	36.8%
Taxable Returns	30,626	3.4%	11.4%	6.4%	13.0%	8.3%
1 to 2.5 million	21,635	2.2%	7.6%	6.1%	10.7%	6.0%
2.5 to 5.0 million	5,505	3.6%	15.1%	7.1%	13.4%	10.4%
5.0 to 10.0 million	2,157	7.6%	22.4%	6.4%	20.7%	15.3%
10.0 to 20 million	824	13.8%	33.3%	7.9%	30.3%	22.3%
over 20.0 million	505	17.8%	46.1%	12.1%	41.8%	34.1%

Source: Internal Revenue Service, Statistics of Income, *Estate Tax Returns Filed in 2003*, IRS, SOI unpublished data, October 2004.

Evidence suggests that most of these estates could still pay the estate tax out of liquid assets alone (cash, bonds, and publicly traded stock) under pre-2001 law, and thus would not have to cash in any other property (such as a home or other real estate). A recent study by the Congressional Budget Office[9] (hereafter the CBO study), examined data from 2000 to determine the fraction of estates that would not have enough liquid assets (stocks, bonds, cash, etc) to pay the tax. (They also indicated that liquid assets would be somewhat higher because it was not able to include assets held in trusts in this study.) For farms, it found that, in 2000, 8% of farm estates could not pay the tax out of liquid assets. Given that an estimated 5% of farm estates pay the tax, less than one half of one percent of farm decedents would be faced with selling any non-liquid assets (whether business or other) to pay the estate tax.

The CBO study did not examine business owners broadly, since it restricted its examination to those business returns that were eligible for QFOBI, and which, therefore, by definition, have estates where business assets account for over half of the estate. These estates are thus much less likely than the average business owner to have sufficient liquid assets to pay the estate tax. But even among these estates, only 32% did not have sufficient liquid assets to pay the estate tax. Thus, it seems likely that the shares of businesses overall that could not pay the tax out of liquid assets are similar in magnitude, and probably smaller, than farms.

Table 2. Percentage of Estate Value Held in Business Assets by Type of Asset and Estate Size in 2003

Size of Gross Estate	Amount (Thousands of Dollars)	Real Estate Partnerships	Closely Held Stock	Farm Assets	Limited Partnerships	Other Non-corp. Bus. Assets
All Returns	194,555,081	1.32%	5.39%	0.44%	1.99%	1.59%
1 to 2.5 million	74,007,063	0.53%	2.34%	0.58%	0.52%	0.62%
2.5 to 5.0 million	35,954,444	0.97%	3.95%	0.44%	1.37%	1.08%
5.0 to 10.0 million	25,285,191	1.18%	5.47%	0.40%	2.28%	1.60%
10.0 to 20 million	17,645,262	2.08%	7.83%	0.39%	2.66%	1.65%
over 20.0 million	41,663,121	2.78%	10.96%	0.23%	4.68%	3.73%
Taxable Returns	109,867,168	1.30%	4.84%	0.30%	2.02%	1.47%
1 to 2.5 million	33,754,841	0.56%	1.37%	0.36%	0.54%	0.36%
2.5 to 5.0 million	18,875,602	0.67%	2.71%	0.35%	1.29%	0.57%
5.0 to 10.0 million	14,684,938	0.72%	3.79%	0.18%	2.10%	1.37%
10.0 to 20 million	11,218,994	1.79%	6.51%	0.27%	2.05%	1.00%
over 20.0 million	31,332,793	2.57%	9.77%	0.26%	4.00%	3.41%

Source: Internal Revenue Service, Statistics of Income, *Estate Tax Returns Filed in 2003*, IRS, SOI unpublished data, October 2004.

These estimates suggest that only a tiny fraction of family-owned businesses (less than ½ of 1%) do not have enough readily available resources to pay the estate tax. Thus, while the estate tax may be a burden on those families, the problem is confined to a small group. In addition, businesses that do not have other assets sufficient to pay the tax still have the option of paying in installments, borrowing, or selling a partial interest in the business. On average these businesses also received significant minority discounts, so that the estate tax owed would be smaller relative to the value of the property.[10]

POLICY OPTIONS FOR ADDRESSING FAMILY BUSINESS ISSUES

There are several alternative policy options for addressing family business estate tax issues. If EGTRRA sunsets, the QFOBI deduction will once again become relevant: the first section below discusses general issues surrounding this provision. The following section discusses the implications for family businesses of making the estate tax repeal permanent. The remaining sections address various revisions within the framework of retaining an estate tax, including an increase in either the QFOBI or general exemption, an unlimited business assets deduction, and an alternative business deduction that would target illiquidity but address a problem with QFOBI due to a "cliff" effect.

Permanent Effects of the Current Law (No Legislative Change and a Reversion to Pre-2001 Law in 2011)

If no legislative action is taken, the estate tax exemption will rise through 2009 (to $3.5 million) and the rate will fall to 45%. In 2010, the estate tax will be repealed, but in 2011 the rules will revert to those that existed before EGTRRA. Under these rules, which were last revised in 1997, the general estate tax exemption will be $1 million, the tax rate will be 55%,[11] and a QFOBI exemption will be allowed, but the combined exemption will still be limited to $1.3 million. Thus, the effective QFOBI exemption is capped at $300,000.

Returning to the status quo and QFOBI raises several issues, a key one being that the real value of the cap has fallen compared to pre-2001 law, so more qualifying estates will be subject to tax. Thus, it might be appropriate to increase the overall cap.

The QFOBI provisions are subject to a number of restrictions designed to ensure that the business is family owned (the decedent and decedent's family must own 50% of the business, or own 30% of a business that is 70% owned by two families, or 90% owned by three families). The business cannot have been publicly traded in the last three years, no more than 35% of the business income may be personal holding company income, and the business must have been owned and operated for five of the past eight years. Heirs are required to continue the business for the next 10 years to avoid recapture. Finally, QFOBI applies only to estates where at least 50% of the assets are family business assets. There are also restrictions on the amount of working capital which are designed to prevent the movement of cash and other liquid assets into the business to increase the deduction or allow an estate to qualify.

The rules designed to target the QFOBI provisions have been criticized due to their complexity. For example, Michael Graetz and Ian Shapiro write:

> Everyone now agrees — regardless of which side of the issue they are on — that QFOBI has been a complete and utter failure... It did not solve anything. QFOBI has so many requirements, so many structures and pitfalls, that very few family businesses have obtained any tax relief at all because of it.[12]

This is a harsh criticism, and it is not necessarily supported by the data. According to the CBO study, for 2000, about 1% of taxable returns claimed the QFOBI and about 1.4% of estates filing returns did so. Given the evidence that only about 3% of taxable estates had more than half of assets in business (and recognizing that some of these estates become nontaxable because of QFOBI), it appears that about a third to a half of these businesses qualified for QFOBI. QFOBI's scope was limited primarily because it was allowed only for those estates where a majority of assets were in the business. Other estates with more than half of their assets in business assets might not have qualified because the business assets were not fully a family business (i.e. largely owned by no more than three families), or because the heirs chose to sell the business.

The QFOBI provision does, nevertheless, raise questions of equity and can produce some economic distortions, as well as complications.

The special business deduction does create some issues of equity — as is the case with any type of special tax exemption. Those with eligible business assets will pay lower taxes on the same amount of wealth than those without such assets. This inequity is a necessary price of targeting a specific group, and must be weighed against the benefits of the QFOBI targeted tax benefit generating a smaller revenue loss than a broader exemption.

QFOBI also creates economic distortions. It induces taxpayers to shift assets into business form, or, alternatively, to avoid liquidating a business when that outcome would be most desirable. How these incentives work depends, in part, on the kind of exemption. Consider an unlimited exemption for business assets. For taxable estates and a 55% estate tax rate, a dollar shifted from non-business assets to business assets saves 55 cents. Or, from another perspective, if business assets earned the same return as other assets, one would be willing to pay $2.22 ($1/(1-0.55)) for a business asset that normally sells for a dollar. Or if asset prices are fixed, one would be willing to accept a rate of return that was as much as 55% smaller in a business investment than in other investments.

The QFOBI provision was limited to estates with 50% or more of gross estate value in business assets, and was capped. Thus, very large estates that had already used up the cap would have no incentive to shift assets, as would estates with so few business assets they would be unlikely to qualify. However smaller estates that fell below the cap and were eligible would have the same incentives as described above for the unlimited simple exemptions (each dollar shifted would, with a 55% tax rate, save 55 cents). And estates that were very close to eligibility would have powerful incentives to shift business investments, though a "cliff" effect. To use a simple example, from pre-2001 law (a $675,000 regular exemption and a maximum of $1.3 million with the business deduction, but using a flat rate for illustrative purposes), if business assets were 45% and the estate totaled $3 million, by shifting $150,000 in assets from non-business to business would save $343,750 (0.55 times $625,000). Even if the assets were virtually worthless, as long as they were not shifted for a long time (causing forgone earnings), one would be better off making the investment. Cliffs are generally to be avoided in devising minimally distorting tax rules.

A third response to this cliff effect would be to overvalue business assets; these rules also magnify existing incentives to undervalue other assets. This response does not distort investment but it does use up resources in estate planning and causes unintended benefits and revenue losses. The cliff effect can also intensify the existing incentives to remove non-business assets from the estate. All of the special business deduction provisions create incentives to recharacterize as much of the estate as possible as business assets.

The provision of QFOBI that required keeping the business in the family after the decedent's death (that recaptured all or part of the tax when the business was ended) also produces economic distortions. The provision affects the allocation of capital and the employment of the heirs. This rule, of course, was intended to target those family businesses whose failure to continue was due to the estate tax.

Finally, the QFOBI provisions complicate administration and compliance. For example, there are rules to prevent holding of cash and other liquid assets in the business; since all businesses must have some cash or near-cash assets available, it is necessary to determine what level of working capital is needed in the business. Such a provision would be necessary with any form of business exemption, since without such rules non-business assets could be lodged in the business. There are also provisions to deal with ownership and control by the family, and whether someone has materially participated in the business.

From this analysis, it seems clear that QFOBI went a long way towards achieving its objective, but it does reflect problems that arose as a result of the targeting of the provision. These problems do not mean that QFOBI was not desirable, as it reflects a specific trade-off of the benefits of addressing the liquidity problem at a minimal revenue cost against the efficiency, equity, and administrative costs. The difficulties in qualifying for QFOBI were perhaps largely because the provision was targeted at family-owned businesses that dominated the estate, not estates that were less clearly "family-owned."

Effect of Making Estate Tax Repeal Permanent

If the estate tax repeal is made permanent, the liquidity issue will disappear for years 2010 and after. However, given that less than 3% of estates could encounter a liquidity problem (returns with more than half of assets in the business, 1.4% for farms and 1.6% for other businesses), this repeal would be very costly from the federal budget perspective if the only objective were to deal with the family-owned business issue.

There is no liquidity problem with this tax regime, because there would be no estate tax and the capital gains tax would apply only if the assets are sold.

This new regime would, however, raise some different issues regarding the effects on the family business. Theoretically, under an income tax, "all accretions to wealth" over a given period of time should be taxed. Traditionally the U.S. income tax has not imposed a tax on gain until realized, and has allowed stepped up basis for appreciated assets so that no tax on the gain accumulated by the decedent is taxed on sale by the heirs. The estate tax provided a backstop to this exemption so that large accumulations of gain would be taxed under the estate tax. The Economic Growth and Tax Relief Reconciliation Act of 2001 (EGTRRA) addressed this issue through changing the asset valuation to a "carry-over" basis regime when the estate tax is repealed in 2010. Under carry-over basis, the heir assumes the basis of the decedent. In other words, the heir "steps into the shoes" of the decedent and would pay capital gains taxes (upon sale of the asset) based on the appreciation from the original purchase price (or value).[13] Congress included in EGTRRA a $3 million spousal exclusion and a $1.3 million general exclusion for capital gains on transferred assets to reduce the tax burden on heirs who sell inherited assets.

Family businesses, and small businesses more generally, likely have significant unrealized capital gains when the proprietor dies. One study estimated the amount of unrealized capital gains held at death for estates valued under $5 million to be approximately 35% of the estates' value (approximately $34 billion in the aggregate in 1998).[14] Any business with total assets valued less than the exemption amount, which is $1.5 million in 2005 rising to $3.5 million in 2009, would not pay any federal estate taxes.[15] In addition, the value of all assets in the business would be stepped up to the value at the time of death. Based on the above estimate of the portion of estate value represented by untaxed capital gains, the step-up treatment confers a significant tax benefit to these estates. The estate tax regime with the stepped-up basis on transferred assets clearly favors these relatively small business estates, sheltering a total of approximately $34 billion in capital gains from taxation. Some have noted that farms in particular benefit from this treatment because, among other reasons, "...the income tax basis of raised animals, for farmers on the cash method of accounting, is zero;...."[16]

When the estate tax is repealed in 2010, the stepped-up basis is replaced with carry-over basis treatment (explained earlier). Congress, however, included a safe harbor for $1.3 million in capital gains passed to a non-spouse. Using the 35% unrealized capital gains estimate above, this treatment would, on average, shelter an estate valued at approximately $3.7 million from capital gains tax liability (if heirs sold bequeathed assets).[17] The smaller the portion of unrealized capital gains, the larger the overall estate size that could avoid capital gains taxation. The difficulty in assessing and confirming the decedent's basis would create an incentive to overstate basis to avoid capital gains taxes.

Theory suggests that the potential capital gains tax liability for the very large estates may generate dynastic asset hoarding. For the largest estates in the NBER study cited above — those with assets over $10 million — the unrealized capital gains comprised over 56% of the estate's total value. The unrealized gains are predominantly business assets in these very large estates; 72.3% of the unrealized gain is held in active farm or business assets. Over time, the potential tax liability would likely increase along with unrealized capital gains further reducing the probability of capital gain realization.

For family businesses that heirs wish to continue, the potential tax liability and asset liquidity are not primary concerns under the carry-over basis regime. In contrast, for heirs who wish to liquidate the business, in particular business with significant untaxed capital gains, the carry-over basis treatment may generate significant tax implications. The disincentive would contribute to a "lock-in" effect, likely growing with each generation.

Increasing the QFOBI Dollar Limits

With no legislative changes, the effective QFOBI exemption has fallen, due to the increase in the general estate tax exemption. The QFOBI dollar limit could be increased, which would reduce the number of firms that are taxable The CBO study provides some estimates of the effects of a general exemption that can also be used to infer some of the effects of raising the combined QFOBI and general exemption. In 2000, 485 estates claiming QFOBI paid tax, and 164 of those had insufficient liquidity to pay the tax. With a $1 million general exemption, a rise to a $1.5 million total (allowing a $500,000 QFOBI deduction) would reduce those numbers by about 50% — to 223 and 82 respectively. Allowing a $2 million total limit (a $1 million QFOBI) would reduce the numbers by about around two thirds — to 135 and 62 respectively.

Retaining the Estate Tax and Expanding the General Exemption and/or Lowering Rates

Another option is to retain the estate tax and stepped-up basis, but increase the general exemption, lower the rates, or both. This approach is much more costly than increasing the QFOBI exemption since it would apply to all estates. But it would alter the taxation of both business and non-business estates substantially. In 2000, 52,000 estates owed taxes and 2,834 (5.5%) had insufficient liquid assets to pay the tax. At that time the general estate tax exemption was $675,000. Increasing the exemption to $1.5 million (the 2005 value), $2 million (the 2006-2008 level), and $3.5 million (the 2009 level) would reduce the number of taxable estates by 70%, 88% and 93% respectively. For all levels of exemptions, between 5 and 6% of taxable estates would not have enough liquid assets to pay the tax. These results suggest that most estates are in the lower part of the asset value distribution, and that large estates have significant liquid assets. In the case of farmers, the proportional reduction would be greater: 82%, 92%, and 96% respectively. The share (although not the number) of estates that would have insufficient liquidity to pay the tax would rise from 8% to 9%, 12%, and 20% respectively. These values suggest that the share of liquid assets in farm estates becomes smaller as estates become larger.

In a different study, the Tax Policy center found that in 2011, with a $1 million exemption, there would be 760 taxable estates with farm or businesses comprising more than 50% of assets; with a $2 million exemption there would be only 210, and with a $3.5 million exemption only 50.[18] These reductions in the tax burden on business assets are, however, accompanied by even more important general reductions in estate taxes, so that, if the objective is to provide relief for family businesses, it is not target efficient.

Clearly an increase in the general exemption would dramatically reduce the number of taxable estates with family-owned businesses.

A General Exemption for Business Property

An alternative approach to raising the general exemption would be to provide an exemption for all business property. With all business property deducted, no estate would have taxes that required the liquidation of the business. With a tax rate of around 50%, about one quarter of the estate would be used to pay taxes if business assets were 50% of the total. Since farm and business assets are slightly over 10% of assets in taxable estates, the cost of this provision would be slightly over 10% of the cost of repealing the estate tax altogether.

An open-ended business exemption would eliminate the cliff effects of the current QFOBI, since all estates with any business property would be eligible to exclude business assets. But, it would also increase the number of estates that would claim a special deduction and increase the amount of assets that receive special tax treatment. The increase in the number of estates and the amount of assets receiving favorable tax treatment would also create distorting effects, encouraging disproportionate investment in business assets across a broad range of estates.

An open ended deduction would also be much more costly than the current QFOBI deductions and would much less precisely target family-owned businesses. Recall that about 40% of taxable estates have business or farm assets, but only about 3% have more than half their assets in business assets, and only about 1% qualify for QFOBI. In addition to expanding the scope of coverage, estates would receive a much larger deduction with an unlimited business exemption.

A general business exemption could still have a dollar cap, which would reduce the cost, relative to an open-ended deduction. Any type of deduction aimed at certain assets would still require rules to prevent abuse (such as limits on cash transferred into the business) and would still create inequities and incentives, although there would no longer be incentives to overvalue business assets or undervalue other assets to qualify for the deduction, as in the case of the cliff in QFOBI.

AN EXEMPTION FOR BUSINESS PROPERTY TARGETED TO LIQUIDITY

Another option is to target the liquidity issue specifically by establishing a business exemption based on the availability of non-business assets to pay the tax. In this case, the business exemption would be a function of both the size of the estate and the share of the estate that is held in business assets. This targeted approach might provide significant revenue savings relative to an exemption that simply excludes all business assets. This approach is illustrated using a simplified example with a flat rate estate tax rate. (Recall that the federal estate tax before the EGTRRA 2001 changes included graduated rates, and the exemption was actually a credit for the taxes that would have been due on the amount up to the exemption amount. The graduated rates (if retained in a final law), however, would need to be taken into consideration under the option described here.)

Congress might wish to consider the policy option of limiting the business exemption so that the tax is no more than the liquidated value of the non-business assets. The reasoning behind this policy objective is readily apparent. Optimally, the estate tax should not force a family business (or farm) to liquidate business assets that generate income, particularly if the business is to continue in the family after the original owner dies. In this scenario, the tax on the entire estate should match the value (at most) of non-business assets.

For example, consider an estate where the total value of all assets in the estate is $10 million, the regular exemption is $1 million, and assume the tax rate is a flat 55% (a simplified version of the pre-2001 tax law rules). Further assume that three-quarters (75%) of the value is held in business assets. If all business assets were exempted, there would be an exemption of $8.5 million ($7.5 million for the business assets and an additional $1 million regular exemption) and the tax due would be $825,000. Suppose instead that an exemption is set precisely so that the tax equals the liquidated value of the all non-business assets (or $2.5 million). This outcome requires a total exemption of $5,454,545 ($2.5 million divided by the tax rate of 55% plus the $1 million regular exemption). Or, stated differently, the tax on what is left in the estate after the business exemption and the regular exemption amount must generate $2.5 million in taxes at the 55% rate.

The estate tax is 55% times ($10,000,000 — $1,000,000 — $4,454,545), or $2.5 million, which is exactly the value of non-business assets. In this case, the effective tax rate is 25% ($2.5 million divided by $10 million). No business faces a liquidity problem, there is no cliff, and the revenue cost of the exemption is less than what would be the case if all business assets were exempt from the estate tax.

In contrast, if exemptions are allowed for all estates, the revenue loss is larger than necessary to target the liquidity issue. If estates are excluded based on a fixed share of business assets (a "cliff" at 50% of total asset value in business assets for example), as in the QFOBI, the rules create very powerful incentives to shift assets to business uses to meet the qualification threshold. This effect will become more powerful if the exemption is unlimited. If the rules precisely target the business assets to deal with the liquidity issue, as described in the previous example, there is also an incentive to shift assets into business uses. In the example provided above, there is an implicit 100% tax on non-business assets, which also creates a powerful incentive to shift assets into the business — not as powerful as a cliff effect, but more powerful than a simple exemption.

For that reason policy makers could opt to modify the formula such that the tax does not consume the entire value of the non-business assets. For example, if every dollar shifted into a business asset increased tax liability by less than a dollar, the incentive effect to reclassify assets is muted. The equation below allows a variety of changes in taxable portion of non-business assets.[19] The variable "*d*" is the portion of the non-business assets that would be used to pay the tax. Reducing *d*, reduces the estate tax burden on non-business assets. If policy makers chose to only devote 60% of non-business assets (and 0% of the business assets) to pay the tax, then *d*=0.60.[20]

$$E^b = A^t - E^s - \frac{A^{nb}d}{t}$$

where,

   t   is the estate tax rate;
   A^t is the total value of all assets in the estate;
   A^{nb} is the total value of non-business/farm assets;
   A^b is the total value of business/farm assets in the estate;
   E^s is standard estate tax exemption amount; and
   E^b is the business/farm asset exemption.

Table 3 illustrates the revenue loss associated with these alternatives for three $10 million estates, assuming a 55% flat rate, and a $1 million general exemption. Note that the difference between the 55% tax and the unlimited business exemption is that the business exemption includes the general exemption when it is smaller, rather than being added to the business exemptions (i.e. each differs by $550,000 which is 55% of $1 million). This adjustment could be made to a general unlimited exemption and would reduce the revenue cost significantly.

Adjusting the exemption of business size in this fashion allows revenue savings for the federal government relative to an unlimited exemption, and eliminates the cliff effect. And, although the equations seem complex, this system would simplify estate planning because qualifying for the exemption would not be as important.

Shifting one dollar from non-business to business use would save the taxpayer less than one dollar, with the size depending on the share parameter (i.e., if the *d* parameter is set at 70%, each dollar shifted saves 70 cents, more than with an unlimited exemption, where it saves 55 cents).

The graphic below exhibits the effective tax rate on a hypothetical estate depending on the business share of assets. If the entire estate is business assets (on the left side of the graph), there is no tax. Each line represents a different value for the policy parameter *d*. Recall that the closer *d* gets to one, the greater the estate tax burden.

Table 3. Revenue Cost of Options for a $10 Million
Estate with a 55% Flat Tax Rate
(revenue cost in $ millions)

Share of Non-business Assets Paid in Tax, the D Parameter	Share of Estate in Business Assets in Three Estates		
	Estate I	Estate II	Estate III
	75%	50%	25%
1.00	2.000	0.000	0.000
0.90	2.250	0.500	0.000
0.80	2.500	0.950	0.000
0.70	2.750	1.450	0.000
0.60	3.000	1.950	0.450
0.55	3.125	2.200	0.825
Unlimited Exemption	3.675	2.750	1.375

Source: Calculations based on equations in the text. The tax rate is 55%; the estate value is $10 million, the value of the standard exemption is $1 million; share of business assets is set at 25%, 50%, and 75%) and *d* ranges from 0.55 to 1.

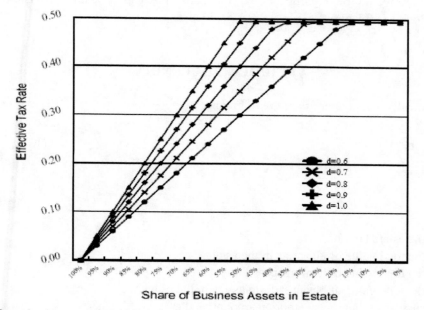

d=share of non-business assets paid in estate tax.

Figure 1. Hypothetical Estate Tax Burden by Share of Business Assets Under Five Policy Parameters.

This type of sliding scale means that the fraction of business assets deducted varies depending on the concentration of business assets in the estate. This outcome may be perceived as unfair; at the same time, the average inequity between those with significant business assets and those with no business assets or only a small share would be reduced.

CONCLUSION

Although the evidence suggests that only a small portion of businesses and farms are subject to the estate tax, there may still be a concern about the impact on those particular businesses.

The options discussed for addressing family-owned business issues involve trade-offs between the revenue cost of providing exemptions to a broader group than necessary to address the issue (including repealing the estate tax in general) and other issues. The more targeted the proposal, the more administrative, distorting, and equity issues arise, but the smaller the revenue cost.

The QFOBI deduction, which is the most targeted of the options considered in this paper, created a particularly difficult problem due to its cliff effect, and the final approach discussed is aimed at eliminating the cliff effect and providing a sliding scale deduction targeted at liquidity issues. Of course, if policy makers desire to eliminate the estate tax as provided in 2010, and there are arguments for doing so, the liquidity issue would disappear. But the revenue effect would be significant and the substitute capital gains provisions would exacerbate lock-in effects.

TECHNICAL APPENDIX

Algebraically, the equations are the following:

$$t(\tilde{A^t} \tilde{E^s_.} E^b) \square A^{nb} \tag{1}$$

$$A^t \square A^b_. A^{nb} \tag{2}$$

where,

t is the estate tax rate;
A^t is the total value of all assets in the estate;
A^{nb} is the total value of non-business/farm assets;
A^b is the total value of business/farm assets in the estate;
E^s is standard estate tax exemption amount; and
E^b is the business/farm asset exemption.

The next step in determining the appropriate value of the business exemption, E^b, is to rearrange equation (1). The following equation, (1)*, for determining the appropriate business exemption if the policy objective is to match the estate tax due to the entire liquidated value of non-business assets.

$$E^b = A^t - E^s - \frac{A^{nb}}{t}$$

(1*)

Equation (3) below allows a variety of changes in taxable portion of nonbusiness assets. The variable "d" is the portion of the non-business assets that would be used to pay the tax. In equation (1)*, $d = 1$, or the total value of non-business assets are used to pay the tax. Reducing d, reduces the estate tax burden on nonbusiness assets. For example, if policy makers chose to only devote 60% of nonbusiness assets (and 0% of the business assets) to pay the tax, then $d = 0.60$.[21]

$$E^b = A^t - E^s - \frac{A^{nb}d}{t}$$

(3)

REFERENCES

[1] For a discussion of the political issues surrounding the estate tax issue, including the issue of family businesses, see Michael J. Graetz and Ian Shapiro, *Death by a Thousand Cuts: The Fight over Taxing Inherited Wealth*, (Princeton: Princeton University Press, 2005).

[2] Non-business assets, such as cash, bonds, and publicly traded stock, are all relatively liquid assets. Under current estate tax rules, the basis of these assets is stepped up to the value at the time of death, thus eliminating the potential for capital gains taxes on inherited appreciation after liquidation.

[3] Some of the deferred tax is subject to a 2% interest payment. The provision allowing current use valuation requires a recapture tax, unless heirs continue to use the land in the business for at least 10 years. The market value can be reduced by a maximum of $750,000 through 1998. After 1998, the maximum is indexed for inflation, rounded to the next lowest multiple of $10,000. In 2004, the maximum was $850,000.

[4] The idea behind the minority discount is that heirs without control are constrained in the flow of income, and the possibility of sale. There are other related discounts, such as marketability discounts (because no ready market exists for the asset). There are concerns that these discounts are excessive in some cases and that they are used through estate planning to minimize estate taxes, and the Joint Tax Committee, in its study *Options to Improve Tax Compliance and Reform Tax Expenditures*, January 27, 2005, has proposed some revisions (see pp. 396-404).

[5] See CRS Report RL32768, Estate and Gift Tax Revenues: Several Measurements, by Nonna Noto.

[6] See CRS Report RL32818, Estate Tax Legislation in the 109th Congress, by Nonna Noto.

[7] See CRS Report RS20593, *Asset Distribution of Taxable Estates*, by Steven Maguire.

[8] CRS Report RL30600, *Estate and Gift Taxes: Economic Issues*, by Jane G. Gravelle and Steven Maguire.

[9] Congressional Budget Office, Effects of the Federal Estate Tax on Farms and Small Businesses, July 2005.

[10] Minority discounts ranged from an average of 16% for residential real estate to 51% for undeveloped land or farmland.

[11] The estate tax actually has graduated rates, but the exemption is effectively a credit, which means that the first dollar taxed is taxed at a rate in the middle of the tax schedule, rather than at the beginning as in the income tax. The first taxable dollar is effectively taxed at 41%, the rates rise to 55%, and there is also a 5% surcharge applying for estates between $10 million and $17.84 million to eliminate the advantages of the lower rate and exemption so that estates above this top limit are taxed at an average rate of 55%.

[12] Graetz and Shapiro, p. 36.

[13] This value is identified as the "cost-basis" of the asset. Ordinarily it is the original purchase price of the asset adjusted for improvements and depreciation in the case of physical assets.

[14] James M. Poterba and Scott Weisbenner, "The Distributional Burden of Taxing Estates and Unrealized Capital Gains at the Time of Death," National Bureau of Economic Research, Working Paper No. 7811, (Cambridge, MA: NBER, July 2000), p. 36.

[15] State estate and inheritance taxes vary by state. Many states have "decoupled" from the federal estate tax and levy stand alone estate and inheritance taxes.

[16] Neil E. Harl, "Taxation of Capital Gains, Gains at Death, and Estates: Policy Considerations," *Tax Notes, Special Report*, July 4, 2005, p. 10.

[17] This estimate is calculated by dividing $1.3 million by 35%.

[18] Reported in Joel Friedman and Ruth Carlitz, "Estate Tax Reform Could Raise Much-Needed Revenue," Center on Budget Policy and Priorities, March 16, 2005.

[19] See the Technical Appendix for the full development of the model.

[20] Also note, that if $d=t$, or the cap is set at the prevailing tax rate, the business exemption is simply the value of business assets less the regular exemption.

[21] Also note, that if $d=t$, or the cap is set at the prevailing tax rate, the business exemption is simply the value of business assets less the regular exemption.

In: Agricultural Finance and Credit
Editor: J. M. Bishoff, pp. 47-59

Chapter 3

CALCULATING ESTATE TAX LIABILITY: 2001 TO 2011 AND BEYOND*

Nonna A. Noto
Specialist in Public Finance Government and Finance Division

ABSTRACT

This report provides a basic explanation of how to calculate the federal estate tax liability for a taxable estate of any given size, using the schedule of marginal tax rates and the applicable exclusion amount or the applicable credit amount for the year of death. The applicable exclusion amount is the amount of any decedent's taxable estate that is free from tax, known informally as the estate tax exemption. The applicable credit amount is the corresponding tax credit equal to the tax that would be due on a taxable estate the size of the applicable exclusion amount.

A shortcut is available to calculate the tax on the estates of decedents dying in 2006 through 2009. The estate tax liability can be calculated simply by multiplying the amount of the taxable estate in excess of the applicable exclusion amount for the year of death times the maximum estate tax rate for the year. The applicable exclusion amount is $2 million for 2006-2008 and $3.5 million for 2009. The maximum tax rate is 46% for 2006 and 45% for 2007-2009.

A more formal method is required to calculate the tax liability for years before 2006 or for 2011 and beyond. This is because more than one marginal tax rate applies to taxable estate values in excess of the exclusion amount. First, the tentative tax that would be due on the entire taxable estate is calculated from the marginal tax rate table for the year of death. Then, the applicable credit amount for the year of death is subtracted from the tentative tax to determine the tax due.

A numerical example is presented in the text and in worksheets for a $5 million taxable estate of a decedent dying in 2007 or 2008. Both the shortcut and formal methods are used to calculate the tax liability.

The Economic Growth and Tax Relief Reconciliation Act of 2001 (P.L. 107-16, EGTRRA) gradually lowered the maximum estate tax rate and substantially raised the applicable exclusion amount over the years 2002 through 2009. The maximum tax rate

---

* Excerpted from CRS Report RL33718, dated November 3, 2006.

fell from 60% under prior law in 2001 (a 55% marginal rate on taxable estate values over $3 million plus a 5% surtax from $10 million to $17 million) to 45% in 2007-2009. The applicable exclusion amount rose from $675,000 in 2001, in steps, up to $3.5 million in 2009. EGTRRA repealed the estate tax for decedents dying in 2010. However, all of the provisions under the act are scheduled to sunset on December 31, 2010. Unless changed beforehand, in 2011 the law will revert back to what it would have been had EGTRRA never been enacted. Under a provision of the Taxpayer Relief Act of 1997 (P.L. 105-34), in 2011 and beyond the applicable exclusion amount would be $1 million, in contrast to $675,000 in 2001. The tables in the appendix present the marginal estate tax rates and the applicable exclusion amount for each year. This report will be updated when there are changes in the law governing estate taxes.

INTRODUCTION

This report provides a basic explanation of how to calculate the federal estate tax liability on a taxable estate of any given size. It uses the schedule of marginal estate tax rates and the applicable exclusion amount, as provided in the Internal Revenue Code, and the corresponding applicable credit amount for the year of death.

The report does not address how to determine the value of the taxable estate except to note that deductions from the gross estate are permitted for funeral expenses, costs of administering the estate, debts and mortgages owed, charitable bequests, and bequests to the surviving spouse. Nor does the report explain the accompanying tax on gifts made during a person's lifetime or the generation-skipping transfer tax.[1]

The Economic Growth and Tax Relief Reconciliation Act of 2001 (P.L. 107-16, EGTRRA, pronounced egg-tra) phased out the federal estate tax over the years 2002-2009. EGTRRA lowered the top marginal estate tax rate gradually and raised the applicable exclusion amount (and consequently the corresponding applicable credit amount) in large increments over the phasedown period. EGTRRA repealed the estate tax for decedents dying in 2010. But the estate tax repeal, and all other provisions of EGTRRA, are scheduled to sunset on December 31, 2010.[2] Unless changed beforehand, in 2011 the law will revert to what it would have been had EGTRRA never been enacted.

APPLICABLE EXCLUSION AMOUNT
AND APPLICABLE CREDIT AMOUNT

It is common parlance to say that a certain amount of a decedent's assets is "free from," "excluded from," or "exempt from" the estate tax. However, the "applicable exclusion amount" is not subtracted from the value of the (net) taxable estate before applying the schedule of tax rates, as a true exemption or deduction would be. Instead, the exclusion amount is converted into a tax credit known as the "applicable credit amount." The credit is equal to the tax that would be due on a taxable estate the size of the applicable exclusion amount for the year of death. This tax credit is subtracted from the "tentative tax" liability on the entire taxable estate.

Unlike a deduction, which is worth more to taxpayers in higher marginal tax rate brackets, the applicable credit amount is worth the same dollar amount regardless of the estate's top tax brackets. However, the credit is non-refundable. That is, it cannot reduce estate tax liability below zero. Consequently, taxable estates that are smaller than the applicable exclusion amount are not able to take full advantage of the available credit.

The applicable exclusion amount also serves as the tax filing threshold. An estate tax return must be filed if the *gross* value of the estate exceeds the applicable exclusion amount for the year of death. This filing requirement holds even if the estate's taxable value falls below this amount after subtracting eligible deductions, such that no estate tax is owed.[3]

Table 1 presents the applicable exclusion amount and the corresponding applicable credit amount for each year from 2001 to 2011 and beyond. In 2001, before EGTRRA, the applicable exclusion amount was $675,000. It was scheduled to rise gradually to $1 million for 2006 and beyond, under provisions of the Taxpayer Relief Act of 1997 (P.L. 105-34, TRA) . EGTRRA raised the applicable exclusion amount to $1 million immediately for 2002 and 2003. It raised the amount further to $1.5 million for 2004 and 2005, $2 million for 2006-2008, and finally $3.5 million for 2009. It repealed the estate tax for 2010 only. If the sunset provision of EGTRRA is not repealed, or the law is not otherwise changed beforehand, in 2011 estate tax law will return to what it would have been had EGTRRA not been enacted. Under the provisions of TRA of 1997, that would mean an applicable exclusion amount of $1 million for the years 2011 and beyond.

The applicable credit amount was calculated by CRS by using the estate tax rate schedule to determine the tentative tax on a taxable estate the size of the applicable exclusion amount. It is based on the underlying graduated marginal tax rates. For the estates of decedents dying in 2002 or 2003, the applicable credit amount was $345,800, corresponding to the estate tax on $1 million. For 2004-2005, the applicable credit amount was $555,800, corresponding to the estate tax on $1.5 million. For 2006-2008, the applicable credit amount is $780,800, corresponding to the estate tax on $2 million. For 2009, the applicable credit amount will be $1,455,800, corresponding to the estate tax on $3.5 million. For 2011 and beyond, the applicable credit amount is scheduled to return to $345,800, the amount that was in effect for 2002-2003, corresponding to the estate tax on $1 million.

Table 1. Applicable Exclusion Amount and Corresponding Applicable Credit Amount, 2001 to 2011 and Beyond

Calendar year (Estates of decedent's death)	Applicable exclusion amount	Applicable credit amount (Equal to the tentative estate tax on the corresponding applicable exclusion amount)
2001	$675,000	$220,550
2002-2003	1,000,000	345,800
2004-2005	1,500,000	555,800
2006-2008	2,000,000	780,800
2009	3,500,000	1,455,800
2010	Estate tax repealed	
2011 and beyond	1,000,000	345,800

Sources: The applicable exclusion amount is from Section 2010(c) of the Internal Revenue Code. The applicable credit amount was calculated by CRS.

MARGINAL TAX RATES

Appendix A presents the graduated schedule of federal estate and gift tax rates for each year from 2001 through 2009. The example here in the text uses the rates that apply to a decedent dying in 2007 or 2008, shown in table A.7. Rising marginal tax rates apply to different portions of an estate. For 2007 and 2008, the statutory rates range from 18% for the taxable value of estates below $10,000, up to 45% for taxable value over $1.5 million.

In practice for 2007 and 2008, the lowest marginal estate tax rate affecting the change in tax liability associated with deductions from or additions to the estate is not 18%, but the maximum rate of 45%. The effect of the applicable exclusion amount is emphasized in italics for over all but the last row of table A.7. These rows encompass taxable estate values from $0 to $1.5 million, and the corresponding statutory marginal tax rates from 18% to 43%. An explanatory row is inserted at the bottom of the table to indicate that the applicable exclusion amount for 2007 and 2008 of $2 million falls within the maximum rate bracket.

CALCULATING ESTATE TAX LIABILITY

A shortcut is available to calculate estate tax liabilities for the estates of decedents dying in 2006 through 2009. This is because the applicable exclusion amounts of $2 million for 2006-2008 and $3.5 million for 2009 fall in the top marginal tax rate bracket. This means that the estate tax liability for 2006-2009 can be calculated simply by multiplying the amount of the taxable estate in excess of the applicable exclusion amount for the year of death times the maximum estate tax rate for the year. For 2006, the applicable exclusion amount was $2 million and the maximum tax rate was 46%. For both 2007 and 2008, the applicable exclusion amount remains at $2 million, but the maximum tax rate falls to 45%. For 2009, the applicable exclusion amount rises to $3.5 million, while the maximum tax rate remains at 45%.

Table 2 illustrates the simplified method of calculating the estate tax liability for a taxable estate of $5 million of a decedent dying in 2007 or 2008. First, subtract the applicable exclusion amount of $2 million from the taxable estate value of $5 million, yielding a difference of $3 million. Then, multiply that excess of $3 million times 0.45 to determine the tax liability of $1.35 million. That represents an average tax rate of 27% ($1.35 million/$5.0 million).

Table 2. Simplified Calculation of Estate Tax Liability for a Taxable Estate of $5,000,000 of a Decedent Dying in 2007 or 2008

Line	Instruction	Estate value, marginal tax rate, and tax liability
1.	Enter value of taxable estate.	$5,000,000
2.	Enter value of applicable exclusion amount for the year of death	$2,000,000
3.	Calculate amount of estate in excess of applicable exclusion amount (line 1 - line 2). Do not enter less than zero.	$3,000,000
4.	Enter the maximum marginal tax rate for the year of death.	0.45
5.	Calculate the estate tax liability (line 3 x line 4).	$1,350,000

Table 3 provides a worksheet for the reader to make a simplified calculation of the estate tax liability of a decedent dying in 2006 through 2009.

Table 3. Worksheet for Simplified Calculation of Estate
Tax Liability for a Decedent Dying in 2006 through 2009

Line	Instruction	Estate value, marginal tax rate, and tax liability
1.	Enter value of taxable estate.	
2.	Enter value of applicable exclusion amount for year of death.	
3.	Calculate amount of estate in excess of applicable exclusion amount (line 1 - line 2). Do not enter less than zero.	
4.	Enter the maximum marginal tax rate for the year of death.	
5.	Calculate the estate tax liability (line 3 x line 4).	

The steps involved in formally calculating the net estate tax liability on a taxable estate for any recent year are laid out in the worksheets shown in tables 4 and 5. Table 4 uses the same numerical example used in table 2 of an estate with a taxable value of $5 million for a decedent dying in 2007 or 2008. Table 5 is a blank worksheet provided for readers to make their own calculations. The formal method set forth in tables 4 and 5 is needed to compute estate tax liability for years before 2006, and for 2011 and beyond. For those years, the marginal tax rate continues to rise above the bracket containing the applicable exclusion amount. For 2006 through 2009, the formal method produces the same tax liability as the shortcut, but requires more computations.

Two basic steps are involved in calculating the estate tax liability. First, calculate the "tentative tax" due on a taxable estate of the size in question, based on the tax rate schedule in the appropriate table in Appendix A for the year of death. Second, subtract from the tentative tax the applicable credit amount for the year of the decedent's death, shown in the last column of table 1.

Calculating the tentative tax, in turn, includes four steps: (1) find in the tax rate table the amount of tax due at the bottom of the bracket in which the estate value falls; (2) calculate the amount of estate in excess of the bottom of the bracket; (3) multiply that excess amount by the marginal tax rate to determined the additional tax; and (4) add that additional tax to the tax due at the bottom of the bracket.

In table 4 the value of the taxable estate — $5 million — is entered on line 1, column (A). To determine the tentative tax on the estate of a decedent dying in 2007 or 2008, look down the first column of table A.7 and find the bracket that encompasses the taxable value of the estate. For an estate of $5 million, that is the top bracket beginning with $1.5 million. Column (c) of the last row of table A.7 shows that the tax on $1.5 million is $555,800. This equals the sum of the taxes due from all of the preceding brackets in the table. Enter this amount on line 2, column (B).

Next, calculate the amount of the estate in excess of $1.5 million, the bottom of the tax rate bracket. In column (A) subtract line 2 from line 1 and enter the difference of $3.5 million on line 3. This $3.5 million is taxed at a marginal rate of 45%, as shown in the last row of table A.7. Enter 0.45 on line 4. Multiply line 4 (0.45) times line 3 ($3,500,000). Enter the resulting additional tax of $1,575,000 on line 5 in column (A) and again in column (B). In column (B), add the additional tax of $1,575,000 on line 5 to the $555,800 on line 2 and enter a total "tentative tax" of $2,130,800 on the $5 million estate on line 6.

Then, find the applicable credit amount corresponding to the year of the decedent's death, in the last column of table 1. For a person dying in 2007 or 2008, the tax credit is $780,800. Enter this amount on line 7. This corresponds to the tax that would be due on an estate of $2 million, the applicable exclusion amount for 2007 or 2008. Subtract the applicable credit amount of $780,800 on line 7 from the tentative tax of $2,130,800 on line 6 and enter the resulting net estate tax liability of $1,350,000 on line 8. This is the same amount arrived at by the simplified method shown in table 2.

Table 4. Formal Calculation of Estate Tax Liability for a Taxable Estate of $5,000,000 of a Decedent Dying in 2007 or 2008

Line	Instruction	(A) Estate value and marginal tax rate	(B) Tax liability ($)
1.	Enter value of taxable estate.	$5,000,000	XX
2.	Find the bracket for this taxable value in the appropriate appendix table A. Enter the bottom of the bracket from table A, col. (a), in col. (A) here. Enter the tax on the bottom of the bracket from table A, col. (c), in col. (B) here.	1,500,000	$555,800
3.	Calculate amount of estate in excess of bottom of bracket: (line 1 - line 2) in col. (A).	3,500,000 (line 1 - line 2)	XX
4.	Enter the marginal tax rate for the bracket, from table A, col. (d), in col. (A) here.	0.45	XX
5.	Calculate tax on amount in excess of bottom of bracket (line 3 x line 4) in col. (A). Enter again in col. (B).	1,575,000 (line 3 x line 4)	1,575,000 (copy)
6.	Calculate tentative tax before credit (line 2 + line 5) in col. (B).	XX	2,130,800 (line 2 + line 5)
7.	Enter applicable credit amount from table 1, for year of death, in col. (B).	XX	780,800 (for 2007 or 2008)
8.	Calculate net estate tax liability (line 6 - line 7) in col. (B). Do not enter less than zero.	XX	1,350,000 (line 6 - line 7)

Notes: XX = intentionally left blank. Author's example in bold, using information from table A.7.

Table 5. Worksheet for Formal Calculation of Estate Tax Liability

Line	Instruction	(A) Estate value ($) and marginal tax rate (%)	(B) Tax liability ($)
1.	Enter value of taxable estate.		XX
2.	Find the bracket for this taxable value in the appropriate appendix table A. Enter the bottom of the bracket from table A, col. (a), in col. (A) here. Enter the tax on the bottom of the bracket from table A, col. (c), in col. (B) here.		
3.	Calculate amount of estate in excess of bottom of bracket: (line 1 - line 2) in col. (A).	(line 1 - line 2)	XX
4.	Enter the marginal tax rate for the bracket, from table A, col. (d), in col. (A) here.		XX
5.	Calculate tax on amount in excess of bottom of bracket (line 3 x line 4) in col. (A). Enter again in col. (B).[a]	(line 3 x line 4)	(copy)
6.	Calculate tentative tax before credit (line 2 + line 5) in col. (B).	XX	(line 2 + line 5)

Table 5. (Continued).

Line	Instruction	(A) Estate value ($) and marginal tax rate (%)	(B) Tax liability ($)
7.	Enter applicable credit amount from table 1, for year of death, in col. (B).	XX	
8.	Calculate net estate tax liability (line 6 - line 7) in col. (B). Do not enter less than zero.	XX	(line 6 - line 7)

Note: XX = No entry needed.

a. On line 5, for 2001 and for 2011and beyond, add a 5% surtax on the taxable value of estates over $10 million up to $17.184 million.

CAUTIONARY NOTES

Readers are warned that this report provides a highly simplified explanation of calculating the estate tax. Other tax credits and deductions can affect the final federal estate tax liability. For example, prior to EGTRRA, an estate could claim a credit (subject to a limit) for state death taxes paid on the estate. EGTRRA phased out the state tax credit by 25 percentage points per year from 2002 through 2004. In 2005, the credit was replaced with a deduction. Because of the sunsetting of EGTRRA, the full state credit is scheduled to be reinstated in 2011. Credits may still be claimed for foreign death taxes and for federal estate taxes paid on property that was transferred to the decedent within the past 10 years (subject to phaseout limits).

From 1998 through 2003, estates with qualified family-owned business interests (QFOBI) could claim a special deduction of up to $675,000 that could, in combination with the applicable exclusion amount, equal $1.3 million. EGTRRA repealed this special deduction as of 2004, when the applicable exclusion amount for all estates reached $1.5 million. The QFOBI deduction would be reinstated in 2011.

Under EGTRRA, the estate and gift taxes remain partly unified from 2002 through 2009. The amount of taxable gifts[4] that may claim the gift tax exclusion is subject to a cumulative lifetime limit of $1 million for 2002 and thereafter. The exclusion claimed for taxable lifetime gifts is subtracted to determine the applicable exclusion amount remaining available to a person's estate at death. The amount of taxable lifetime gifts is included in the base used to calculate the tentative estate tax. However, the estate is credited for gift taxes previously paid. In 2010, the gift tax is scheduled to remain in place when the estate tax is repealed. The lifetime applicable exclusion amount for gifts will remain at $1 million. The maximum rate of tax will be 35% on cumulative gifts over $500,000. Under the return to prior law in 2011, the fully unified estate and gift tax would be reinstated. The single, combined exclusion of $1 million could be applied to gifts and/or bequests.

A separate tax is levied on generation-skipping transfers (GST) through 2009. The GST tax is repealed for 2010. The GST tax would be reinstated in 2011.

Prior to EGTRRA, a 5% surtax was levied on taxable estate values from $10 million through $17.184 million to reclaim the benefits from the graduated marginal rates below 55%. The surtax was repealed in 2002. The surtax would be reinstated in 2011.

APPENDIX A.
FEDERAL ESTATE TAX RATE SCHEDULES FOR 2001 THROUGH 2009 AND 2011 AND BEYOND

During the phasedown period for the estate tax, from 2002 through 2009, the provisions of the Economic Growth and Tax Relief Reconciliation Act of 2001 (P.L. 107-16, EGTRRA) gradually reduce the maximum estate tax rate and substantially raise the applicable exclusion amount. The tables in this appendix present the changes year-by-year. The graduated rates from 18% to 43% established under prior law continue to apply to estate values up to $1.5 million. The maximum estate tax rate dropped from 55% on taxable estate values over $3 million (plus a 5% surtax in a specified range over $10 million) in 2001 under prior law, to 50% on taxable estates over $2.5 million in 2002. The maximum tax rate continued to drop by one percentage point per year, down to 45% on taxable estate values over $1.5 million in 2007-2009. Notes at the bottom of each table summarize any reduction in the maximum estate tax rate or increase in the applicable exclusion amount scheduled to take effect that year.

There is an explanatory row in each table indicating where the applicable exclusion amount for that year of death falls within the rate schedule. Taxable estate values up to the applicable exclusion amount are protected from tax by the unified tax credit for the year. The rows of marginal tax rates corresponding to estate values below the exclusion amount are in italics. The marginal tax rates that affect the changes in tax liability associated with deductions from, or additions to, the taxable estate are those above the exclusion amount. For 2007-2008 and 2009 (tables A.7 and A.8), the applicable exclusion amounts fall within the top-rate bracket of 45%.

The tax rates presented in table A.1 are the rates that were in effect prior to the enactment of EGTRRA. They continued to apply to the estates of decedents who died in 2001. They are also the rates that will be reinstated in 2011 if EGTRRA is allowed to sunset on December 31, 2010. A notable difference is that, under a provision of the Taxpayer Relief Act of 1997 (P.L. 105-34), the applicable exclusion amount will be $1 million for 2011 and beyond, in contrast to $675,000 in 2001.

The estate tax rate schedules for 2002-2009 presented in this appendix were created by CRS. The tax rate schedule under prior law, from Section 2001(c)(1) of the Internal Revenue Code, was adapted according to the information presented in the new Section 2001(c)(2). Section 2001(c)(2) sets forth the reduction in maximum estate tax rates scheduled by EGTRRA.

Table A.1. Federal Estate Tax Rate Schedule for 2001 and for 2011 and Beyond

Taxable Estate		Tentative Tax	
(a) Bottom of bracket	(b) Top of bracket	(c) Tax on bottom of bracket	(d) Marginal tax rate on bracket
Over	But Not Over	Tax on amount in column (a)	Plus: Rate of tax on amount over bottom of bracket, in column (a), up to top of bracket, in column (b)
$0	$10,000	$0 +	18% of such amount
10,000	20,000	1,800 +	20% of excess over $10,000
20,000	40,000	3,800 +	22% of excess over 20,000
40,000	60,000	8,200 +	24% of excess over 40,000
60,000	80,000	13,000 +	26% of excess over 60,000
80,000	100,000	18,200 +	28% of excess over 80,000
100,000	150,000	23,800 +	30% of excess over 100,000
150,000	250,000	38,800 +	32% of excess over 150,000
250,000	500,000	70,800 +	34% of excess over 250,000
Applicable Exclusion Amount was $675,000 for 2001			
500,000	750,000	155,800 +	37% of excess over 500,000
750,000	1,000,000	248,300 +	39% of excess over 750,000
Applicable Exclusion Amount will be $1,000,000 for 2011 and beyond			
1,000,000	1,250,000	345,800 +	41% of excess over 1,000,000
1,250,000	1,500,000	448,300 +	43% of excess over 1,250,000
1,500,000	2,000,000	555,800 +	45% of excess over 1,500,000
2,000,000	2,500,000	780,800 +	49% of excess over 2,000,000
2,500,000	3,000,000	1,025,800 +	53% of excess over 2,500,000
over 3,000,000	—	1,290,800 +	55% of excess over 3,000,000
Special range for 5% surtax to phase out the benefit of graduated tax rates			
10,000,000	17,184,000	Maximum surtax: 5% x $7,184,000 = $359,200	5% surtax means a 60% marginal tax rate in this range

Note: These are the tax rates that were in effect prior to enactment of the Economic Growth and Tax Relief Act of 2001 (P.L. 107-16, EGTRRA).

Table A.2. Federal Estate Tax Rate Schedule for 2002

Taxable Estate		Tentative Tax	
(a) Bottom of bracket	(b) Top of bracket	(c) Tax on bottom of bracket	(d) Marginal tax rate on bracket
Over	But Not Over	Tax on amount in column (a)	Plus: Rate of tax on amount over bottom of bracket, in column (a), up to top of bracket, in column (b)
$0	$10,000	$0 +	18% of such amount
10,000	20,000	1,800 +	20% of excess over $10,000
20,000	40,000	3,800 +	22% of excess over 20,000
40,000	60,000	8,200 +	24% of excess over 40,000
60,000	80,000	13,000 +	26% of excess over 60,000
80,000	100,000	18,200 +	28% of excess over 80,000
100,000	150,000	23,800 +	30% of excess over 100,000
150,000	250,000	38,800 +	32% of excess over 150,000
250,000	500,000	70,800 +	34% of excess over 250,000
500,000	750,000	155,800 +	37% of excess over 500,000
750,000	1,000,000	248,300 +	39% of excess over 750,000

Table A.2. (Continued).

Taxable Estate			Tentative Tax
(a) Bottom of bracket	(b) Top of bracket	(c) Tax on bottom of bracket	(d) Marginal tax rate on bracket
Over	But Not Over	Tax on amount in column (a)	Plus: Rate of tax on amount over bottom of bracket, in column (a), up to top of bracket, in column (b)
Applicable Exclusion Amount was $1 Million for 2002			
1,000,000	1,250,000	345,800 +	41% of excess over 1,000,000
1,250,000	1,500,000	448,300 +	43% of excess over 1,250,000
1,500,000	2,000,000	555,800 +	45% of excess over 1,500,000
2,000,000	2,500,000	780,800 +	49% of excess over 2,000,000
2,500,000	—	1,025,800 +	50% of excess over 2,500,000

Notes: In 2002, the 5% surtax was repealed, as were tax rates in excess of 50%. In 2002, the maximum rate was 50% on taxable amounts over $2.5 million. The applicable exclusion amount rose from $675,000 in 2001 under prior law to $1 million in 2002 under EGTRRA.

Table A.3. Federal Estate Tax Rate Schedule for 2003

Taxable Estate			Tentative Tax
(a) Bottom of bracket	(b) Top of bracket	(c) Tax on bottom of bracket	(d) Marginal tax rate on bracket
Over	But Not Over	Tax on amount in column (a)	Plus: Rate of tax on amount over bottom of bracket, in column (a), up to top of bracket, in column (b)
$ 0	$ 10,000	$0 +	18% of such amount
10,000	20,000	1,800 +	20% of excess over $10,000
20,000	40,000	3,800 +	22% of excess over 20,000
40,000	60,000	8,200 +	24% of excess over 40,000
60,000	80,000	13,000 +	26% of excess over 60,000
80,000	100,000	18,200 +	28% of excess over 80,000
100,000	150,000	23,800 +	30% of excess over 100,000
150,000	250,000	38,800 +	32% of excess over 150,000
250,000	500,000	70,800 +	34% of excess over 250,000
500,000	750,000	155,800 +	37% of excess over 500,000
750,000	1,000,000	248,300 +	39% of excess over 750,000
Applicable Exclusion Amount was $1 Million for 2003			
1,000,000	1,250,000	345,800 +	41% of excess over 1,000,000
1,250,000	1,500,000	448,300 +	43% of excess over 1,250,000
1,500,000	2,000,000	555,800 +	45% of excess over 1,500,000
2,000,000	—	780,800 +	49% of excess over 2,000,000

Notes: In 2003, tax rates above 49% were repealed. The tax rate on taxable estate values over $2.5 million fell from 50% to 49%. The applicable exclusion amount remained at $1 million.

Table A.4. Federal Estate Tax Rate Schedule for 2004

Taxable Estate		Tentative Tax	
(a) Bottom of bracket	(b) Top of bracket	(c) Tax on bottom of bracket	(d) Marginal tax rate on bracket
Over	But Not Over	Tax on amount in column (a)	Plus: Rate of tax on amount over bottom of bracket, in column (a), up to top of bracket, in column (b)
$ 0	$ 10,000	$0 +	18% of such amount
10,000	20,000	1,800 +	20% of excess over $10,000
20,000	40,000	3,800 +	22% of excess over 20,000
40,000	60,000	8,200 +	24% of excess over 40,000
60,000	80,000	13,000 +	26% of excess over 60,000
80,000	100,000	18,200 +	28% of excess over 80,000
100,000	150,000	23,800 +	30% of excess over 100,000
150,000	250,000	38,800 +	32% of excess over 150,000
250,000	500,000	70,800 +	34% of excess over 250,000
500,000	750,000	155,800 +	37% of excess over 500,000
750,000	1,000,000	248,300 +	39% of excess over 750,000
1,000,000	1,250,000	345,800 +	41% of excess over 1,000,000
1,250,000	1,500,000	448,300 +	43% of excess over 1,250,000
Applicable Exclusion Amount was $1.5 Million for 2004			
1,500,000	2,000,000	555,800 +	45% of excess over 1,500,000
2,000,000	—	780,800 +	48% of excess over 2,000,000

Notes: In 2004, tax rates above 48% were repealed. The tax rate on taxable estate values over $2 million fell from 49% to 48%. The applicable exclusion amount rose from $1 million to $1.5 million.

Table A.5. Federal Estate Tax Rate Schedule for 2005

Taxable Estate		Tentative Tax	
(a) Bottom of bracket	(b) Top of bracket	(c) Tax on bottom of bracket	(d) Marginal tax rate on bracket
Over	But Not Over	Tax on amount in column (a)	Plus: Rate of tax on amount over bottom of bracket, in column (a), up to top of bracket, in column (b)
$ 0	$ 10,000	$0 +	18% of such amount
10,000	20,000	1,800 +	20% of excess over $10,000
20,000	40,000	3,800 +	22% of excess over 20,000
40,000	60,000	8,200 +	24% of excess over 40,000
60,000	80,000	13,000 +	26% of excess over 60,000
80,000	100,000	18,200 +	28% of excess over 80,000
100,000	150,000	23,800 +	30% of excess over 100,000
150,000	250,000	38,800 +	32% of excess over 150,000
250,000	500,000	70,800 +	34% of excess over 250,000
500,000	750,000	155,800 +	37% of excess over 500,000
750,000	1,000,000	248,300 +	39% of excess over 750,000
1,000,000	1,250,000	345,800 +	41% of excess over 1,000,000
1,250,000	1,500,000	448,300 +	43% of excess over 1,250,000
Applicable Exclusion Amount was $1.5 Million for 2005			
1,500,000	2,000,000	555,800 +	45% of excess over 1,500,000
2,000,000	—	780,800 +	47% of excess over 2,000,000

Notes: In 2005, tax rates above 47% were repealed. The tax rate on taxable estate values over $2 million fell from 48% to 47%. The applicable exclusion amount remained at $1.5 million.

Table A.6. Federal Estate Tax Rate Schedule for 2006

Taxable Estate		Tentative Tax	
(a) Bottom of bracket	(b) Top of bracket	(c) Tax on bottom of bracket	(d) Marginal tax rate on bracket
Over	But Not Over	Tax on amount in column (a)	Plus: Rate of tax on amount over bottom of bracket, in column (a), up to top of bracket, in column (b)
$ 0	$ 10,000	$0 +	18% of such amount
10,000	20,000	1,800 +	20% of excess over $10,000
20,000	40,000	3,800 +	22% of excess over 20,000
40,000	60,000	8,200 +	24% of excess over 40,000
60,000	80,000	13,000 +	26% of excess over 60,000
80,000	100,000	18,200 +	28% of excess over 80,000
100,000	150,000	23,800 +	30% of excess over 100,000
150,000	250,000	38,800 +	32% of excess over 150,000
250,000	500,000	70,800 +	34% of excess over 250,000
500,000	750,000	155,800 +	37% of excess over 500,000
750,000	1,000,000	248,300 +	39% of excess over 750,000
1,000,000	1,250,000	345,800 +	41% of excess over 1,000,000
1,250,000	1,500,000	448,300 +	43% of excess over 1,250,000
1,500,000	2,000,000	555,800 +	45% of excess over 1,500,000
Applicable Exclusion Amount was $2 Million for 2006			
2,000,000	—	780,800 +	46% of excess over 2,000,000

Notes: In 2006, tax rates above 46% were repealed. The tax rate on taxable estate values over $2 million fell from 47% to 46%. The applicable exclusion amount rose from $1.5 million to $2 million.

Table A.7. Federal Estate Tax Rate Schedule for 2007 and 2008

Taxable Estate		Tentative Tax	
(a) Bottom of bracket	(b) Top of bracket	(c) Tax on bottom of bracket	(d) Marginal tax rate on bracket
Over	But Not Over	Tax on amount in column (a)	Plus: Rate of tax on amount over bottom of bracket, in column (a), up to top of bracket, in column (b)
$0	$ 10,000	$0 +	18% of such amount
10,000	20,000	1,800 +	20% of excess over $10,000
20,000	40,000	3,800 +	22% of excess over 20,000
40,000	60,000	8,200 +	24% of excess over 40,000
60,000	80,000	13,000 +	26% of excess over 60,000
80,000	100,000	18,200 +	28% of excess over 80,000
100,000	150,000	23,800 +	30% of excess over 100,000
150,000	250,000	38,800 +	32% of excess over 150,000
250,000	500,000	70,800 +	34% of excess over 250,000
500,000	750,000	155,800 +	37% of excess over 500,000
1,000,000	1,250,000	345,800 +	41% of excess over 1,000,000
1,250,000	1,500,000	448,300 +	43% of excess over 1,250,000
1,500,000	—	555,800 +	45% of excess over 1,500,000
Applicable Exclusion Amount will remain at $2 Million for 2007 and 2008			

Notes: In 2007, tax rates above 45% will be repealed. The tax rate on taxable estate values over $2 million will be lowered from 46% to 45%. This is the lowest rate provided by EGTRRA. The tax rate was already 45% on taxable estate values from $1.5 million to $2 million, under prior law. Thus, the tax rate will be 45% on taxable estate values over $1.5 million. The applicable exclusion amount will remain at $2 million, as it was in 2006. The law for 2008 will remain the same as for 2007.

Table A.8. Federal Estate Tax Rate Schedule for 2009

Taxable Estate		Tentative Tax	
(a) Bottom of bracket	(b) Top of bracket	(c) Tax on bottom of bracket	(d) Marginal tax rate on bracket
Over	But Not Over	Tax on amount in column (a)	Plus: Rate of tax on amount over bottom of bracket, in column (a), up to top of bracket, in column (b)
$ 0	$ 10,000	$0 +	18% of such amount
10,000	20,000	1,800 +	20% of excess over $10,000
20,000	40,000	3,800 +	22% of excess over 20,000
40,000	60,000	8,200 +	24% of excess over 40,000
60,000	80,000	13,000 +	26% of excess over 60,000
80,000	100,000	18,200 +	28% of excess over 80,000
150,000	250,000	38,800 +	32% of excess over 150,000
250,000	500,000	70,800 +	34% of excess over 250,000
500,000	750,000	155,800 +	37% of excess over 500,000
750,000	1,000,000	248,300 +	39% of excess over 750,000
1,000,000	1,250,000	345,800 +	41% of excess over 1,000,000
1,250,000	1,500,000	448,300 +	43% of excess over 1,250,000
1,500,000	—	555,800 +	45% of excess over 1,500,000
Applicable Exclusion Amount will be $3.5 Million for 2009			

Notes: In 2009, the applicable exclusion amount will rise from $2 million to $3.5 million. There will be no further reduction in the maximum tax rate.

REFERENCES

[1] For more information about these topics, see CRS Report RL30600, *Estate and Gift Taxes: Economic Issues*, by Jane G. Gravelle and Steven Maguire and CRS Report 95-416, *Federal Estate, Gift, and Generation, Skipping Taxes: A Description Of Current Law*, by John R. Luckey.

[2] For a fuller explanation of the estate tax provisions in EGTRRA, see CRS Report RL31061, *Estate and Gift Tax Law: Changes Under the Economic Growth and Tax Relief Reconciliation Act of 2001*, by Nonna A. Noto; and CRS Report RS20989, *Federal Estate, Gift, and Generation-Skipping Transfer Taxes: Modification, Phase Out and Repeal Under the Economic Growth And Tax Relief Reconciliation Act of 2001*, by John R. Luckey.

[3] Less than half of estate tax returns filed in 1998 through 2004 were taxable. For additional information, see CRS Report RL32768, *Estate and Gift Tax Revenues: Several Measurements*, by Nonna A. Noto.

[4] Gift amounts that are protected from taxation by the annual exclusion amount ($12,000 per donor per donee in 2006 and 2007, indexed for inflation,) do not count against the cumulative lifetime exclusion.

In: Agricultural Finance and Credit
Editor: J. M. Bishoff, pp. 61-66

ISBN: 978-1-60456-072-5
© 2008 Nova Science Publishers, Inc.

Chapter 4

CAPITAL GAINS TAXES: AN OVERVIEW*

Jane G. Gravelle

Senior Specialist in Economic Policy Government and Finance Division

ABSTRACT

Tax legislation in 1997 reduced capital gains taxes on several types of assets, imposing a 20% maximum tax rate on long-term gains, a rate temporarily reduced to 15% for 2003-2008, which was extended for two additional years in 2006. There is also an exclusion of $500,000 ($250,000 for single returns) for gains on home sales. The capital gains tax has been a tax cut target since the 1986 Tax Reform Act treated capital gains as ordinary income. An argument for lower capital gains taxes is reduction of the lock-in effect. Some also believe that lower capital gains taxes will cost little compared to the benefits they bring and that lower taxes induce additional economic growth, although the magnitude of these potential effects is in some dispute. Others criticize lower capital gains taxes as benefitting higher income individuals and express concerns about the budget effects, particularly in future years. Another criticism of lower rates is the possible role of a larger capital gains tax differential in encouraging tax sheltering activities and adding complexity to the tax law.

WHAT ARE CAPITAL GAINS AND HOW ARE THEY TAXED?

Capital gain arises when an asset is sold and consists of the difference between the basis (normally the acquisition price) and the sales price. Corporate stock accounts for 20% to 80% of taxable gains, depending on stock market performance. Real estate is the remaining major source of capital gains, although gain also arises from other assets (e.g., timber sales and collectibles). The appreciation in value can be real or reflect inflation. Corporate stock appreciates both because the firm's assets increase with reinvested earnings and because general price levels are rising. Appreciation in the value of property may simply reflect

---

* Excerpted from CRS Report 96-769, dated January 24, 2007.

inflation. For depreciable assets, some of the gain may reflect the possibility that the property was depreciated too quickly.

If the return to capital gains were to be effectively taxed at the statutory tax rate in the manner of other income, real gains would have to be taxed in the year they accrue. Current practice departs from this approach. Gains are not taxed until realized, benefitting from the deferral of taxes. (Taxes on interest income are due as the interest is accrued). Gains on an asset held until death may be passed on to heirs with the tax forgiven; if the asset is then sold, the gain is sales price less market value at the time of death, a treatment referred to as a "step-up in basis."

Under current law, there is a maximum tax of 20% on capital gains held for a year (temporarily reduced to 15% for 2003-2010), although the ordinary income tax rates reach as high as 39.6% (temporarily reduced to 35%). Where ordinary tax rates are 15% or below, the capital gains tax is 10% (temporarily at 5% for 2003-2007 and 0% for 2008-2010). Assuming the temporary provisions expire, gain from assets held five years and acquired after 2000 will be subject to a maximum rate of 18%. For gains in the 15% bracket and below, an 8% rate will apply to any gain on assets held for five years and sold after 2000, with no required acquisition date. Gain arising from prior depreciation deductions is taxed at ordinary rates, but there is a 25% ceiling rate on the gain from attributable to prior straight-line depreciation on real property.

Under law prior to 1997, several rules permitted avoidance or deferral of the tax on gain on owner-occupied housing, including a provision allowing deferral of gain until a subsequent house is sold (rollover treatment) and a provision allowing a one-time exclusion of $125,000 of gain for those 55 and over. These provisions were replaced with a general $500,000 exclusion ($250,000 for a single individual), which cost only slightly more in revenue.

In contrast to these provisions that benefit capital gains, capital gains are penalized because many of the gains that are subject to tax arise from inflation and therefore do not reflect real income.

A BRIEF HISTORY

Capital gains were taxed when the income tax (with rates up to 7%) was imposed in 1913. An alternative rate of 12.5% was allowed in 1921 (the regular top rate was 73%). Tax rates were cut several times during the 1920s. Capital gain exclusions based on holding period were enacted in 1924, and modified in 1938, to deal with bunching of gains in one year. In 1942 a 50% exclusion was adopted, with an alternative rate of 25%. Over time, the top rate on ordinary income varied, rising to 94% in the mid-1940s, then dropping to 70% after 1964. In 1969 a new minimum tax increased the gains tax for some; the 25% alternative tax was repealed.

In 1978 the minimum tax on capital gains was repealed and the exclusion increased to 60% with a maximum rate of 28% (0.4 times 0.7). The top rate on ordinary income was reduced to 50% in 1981, reducing the capital gains rate to 20% (0.4 times 0.5). The Tax Reform Act of 1986 reduced tax rates further, but, in order to maintain distributional neutrality, eliminated some tax preferences, including the exclusion for capital gains. This treatment brought the rate for high income individuals in line with the rate on ordinary income — 28%.

In 1989, President Bush proposed a top rate of 15%, halving top rates. The Ways and Means Committee considered two proposals: Chairman Rostenkowski proposed to index capital gains, and Representatives Jenkins, Flippo, and Archer proposed a 30% capital gains exclusion through 1991 followed by inflation indexation. This latter measure was approved by the Committee, but was not enacted.

In 1990, the President proposed a 30% exclusion, setting the rate at 19.6% for high income individuals. The House also passed a 50% exclusion with a lifetime maximum ceiling and a $1,000 annual exclusion, but this provision was not enacted into law. When rates on high income individuals were set at 31%, however, the capital gains rate was capped at 28%.

In 1991, the President again proposed a 30% exclusion, but no action was taken. In 1992, the President proposed a 45% exclusion. The House adopted a proposal for indexation for inflation for newly acquired assets: the Senate passed a separate set of graduated rates on capital gains that tended to benefit more moderate income individuals. This latter provision was included in a bill (H.R. 4210) containing many other tax provisions that was vetoed by the President.

No changes were proposed by President Clinton or adopted in 1993 and 1994 with the exception of a narrowly targeted benefit for small business stock adopted in 1993. The value of the tax cap was increased, however, in 1993 when new brackets of 36% and 39.6% were added for ordinary income.

In 1994, the "Contract With America" proposed a 50% exclusion for capital gains, and indexing the basis for all subsequent inflation, while eliminating the 28% cap; this exclusion would be about a 40% reduction on average from current rates. The Ways and Means Committee reported out H.R. 1215, which restricted inflation indexing to newly acquired assets (individuals could "mark to market" — pay tax on the difference between fair market value and basis as if the property were sold to qualify for indexation), did not allow indexation to create losses and provided a flat 25% tax rate for corporations). The 1995 reconciliation bill (H.R. 2491) that was vetoed by the President, included these revisions but delayed the indexation provision until 2002. During the 1996 presidential election, Mr. Dole proposed a slightly larger capital gains cut, and both candidates supported elimination of capital gains taxes on virtually all gains from home sales.

In 1997, the President and Congress agreed to a tax cut as part of the reconciliation. The Administration tax cut proposal included the change in tax treatment of owner occupied housing. The House bill included a reduction in the 15% and 28% rates to 10% and 20%, about a 30% cut. Capital gains would also be indexed for assets acquired after 2000 and held for three years; mark-to-market would also be allowed. The Senate and the final bill did not include indexing. The capital gains issue was briefly revisited in 1998, when the holding period for long term gains was moved back from the 18 months set in 1997 to the one year period that has typically applied. The 1999 House bill would have cut the rates to 15% and 10%: the conference version cut rates to 18% and 8% and proposed indexing of future gains, but the bill was vetoed. Capital gains were discussed during the consideration of the economic stimulus bill at the end of 2002, but not included in any legislative proposal (and no proposal was adopted). The temporary provisions for lower rates of 15% for 2003-2008 for those in the higher brackets and to 5% in 2003-2007 and 0% in 2008 for taxpayers in the 15%

bracket or lower were adopted in 2003. H.R. 4297, adopted in 2006, extended these lower rates for two more years.

REVENUE EFFECTS

Over the past several years, a debate has ensued regarding the revenue cost of cutting capital gains taxes (see CRS Report 97-559, *The Revenue Cost of Cutting Capital Gains Tax Rates*, for further discussion). For example, when the President proposed a 30% exclusion in 1990, Treasury estimates showed a $12 billion gain in revenue over the first five years, while the Joint Committee on Taxation found a revenue loss of approximately equal size.

Although the estimates seemed quite different, they both incorporated significant expected increases in the amount of gains realized as a result of the tax cut. For example, the Treasury would have estimated a revenue loss of $80 billion over five years with no behavioral response, and the Joint Tax Committee a loss of $100 billion. (The gap between these static estimates arose from differences in projections of expected capital gains, a volatile series that is quite difficult to estimate.)

Empirical evidence on capital gains realizations does not clearly point to a specific response and revenue cost. Recent research suggests long-run responses may be more modest than those suggested by the economics literature during the 1990 debate, but the short-run response is still difficult to ascertain. (For a survey of this literature, see Jane G. Gravelle, *The Economic Effects of Taxing Capital Income*, Cambridge, MA.: MIT Press, 1984, pp. 143-151; see also Leonard Burman, *The Labyrinth of Capital Gains Tax Policy*, Washington, DC: The Brookings Institution, 1999, for a discussion of this issue and many others.)

Any revenue feedback effect will be smaller the larger the tax reduction. When the tax reduction is large, although there will be a larger response, any induced revenues will be taxed at the new lower rates. Thus, a 50% exclusion will not have as large a feedback effect relative to the static estimate as a 30% exclusion. Moreover, allowing a prospective tax cut that depends on selling and acquiring a new asset to qualify (or marking to market) as is the case for the reduction from 20% to 18% for five-year property causes a gain in the short run as well.

Arguments have also been made that a capital gains tax cut will induce additional savings, also resulting in a feedback effect as taxes are imposed on new income. This effect is uncertain, as it is not clear that an increase in the rate of return will increase savings (savings can decrease if the income effect is more powerful than the substitution effect) and what the magnitude of the response might be. (See Congressional Budget Office Memorandum, *An Analysis of the Potential Macroeconomic Effects of the Economic Growth Act of 1998*). There is also a debate about the effect of the capital gains tax on growth through its effect on innovation. Regardless of these empirical uncertainties, any effect of savings on taxable income in the short run is likely to be quite small due to the slow rate of capital accumulation. (Net savings are typically only about 2% to 3% of the capital stock, so that even a 10% increase in the savings rate would result in only a 2/10 to 3/10 of a percent first-year increase in the capital stock.) A related argument is that the tax cut will increase asset values; such an effect is only temporary, however, and will, if it occurs, only shift revenues from the future to the present. (For a discussion of savings and asset valuations, see testimony of Jane G.

Gravelle, Congressional Research Service before the Senate Finance Committee, February 15, 1995 and the House Ways and Means Committee, March 19, 1997.)

IMPACT ON EFFECTIVE TAX BURDENS

The tax burden on an investment is influenced by both the tax rate and any benefits allowed or penalties imposed. One way to measure this tax burden is to calculate a marginal effective tax rate that captures in a single number all of the factors that affect tax burden. It is the percentage difference between the before- and after-tax return to investment, or the estimated statutory rate that would be applied to economic income to give the taxpayer the same burden as the combination of tax benefits and penalties.

The effective tax rate on capital gains can be either higher or lower than the statutory rate, depending on the inflation rate relative to the real appreciation rate and the holding period. Also, assets held until death are not subject to tax. For example, under prior law, assuming a 20% statutory tax rate, the gain on a growth stock (paying no dividends) with a real appreciation rate of 7% and an inflation rate of 3% would, if held for one year, seven years, 20 years, and until death, be subject respectively to tax rates of 27%, 22%, 14%, and 0%. The rate on 7- and 20-year assets would be 18%, for effective tax rates of 19% and 12%. (With no inflation and a 20% rate, the rates would be 19%, 16%, 12% and 0%; inflation penalizes assets held a shorter period more heavily than assets held for a longer period). Since less than half of gains that are accrued are realized, the effective tax rate is probably lower than the statutory tax rate. These benefits are larger for individuals who in the highest tax brackets (31%, 36%, and 39.6%) because the capital gains tax rate is capped at 20%. The rates would be lowered proportionally with a 15% tax rate.

ISSUES: EFFICIENCY, GROWTH, DISTRIBUTION, AND COMPLEXITY

One argument in favor of reducing the capital gains tax is the lock-in effect. If this effect is large, the tax introduces significant distortions in behavior with relatively little revenue gain. Another way to reduce lock-in is accrual taxation (i.e., tax gains on a current basis as accrued), but this approach is only feasible when assets can be easily valued (e.g., publicly traded corporate stock). Another way to reduce lock-in is to tax gains passed on at death. These solutions may face technical problems and taxation of gains at death has been unpopular.

For owner-occupied housing, although there were many ways to avoid the tax under current law, the rules may have resulted in individuals remaining in houses that are too large if economic circumstances have declined (for example, through job loss or retirement or from a preference for a smaller home, or if there is relocation to a lower-cost area.)

A case might be made for lower capital gains taxes on corporate stock because corporate equity capital is subject to heavy taxation. This heavy taxation encourages corporations to take on too much debt and directs too much capital to the noncorporate sector. On the other hand, lower capital gains taxes increase the relative penalty that applies to dividends and introduce tax distortions in the decisions of the firm to retain earnings. This latter effect, however, does not occur under the temporary tax cuts that benefit dividends as well as capital gains.

Arguments have also been made that lower gains taxes will increase economic growth and entrepreneurship. Although evidence on the effect of tax cuts on savings rates and, thus, economic growth is difficult to obtain, most evidence does not indicate a large response of savings to an increase in the rate of return. Indeed, not all studies find a positive response, because a higher rate of return may allow individuals to save less while reaching their desired goal. (See Gravelle, *The Economics of Taxing Capital Income*, MIT Press, pp. 24-28, for a survey.) A more effective route to increasing savings may be to take revenues that might otherwise finance a tax cut and reduce the debt.

Although arguments are made that lower gains taxes stimulate innovation and entrepreneurship, there is little evidence in history to connect periods of technical advance with lower taxes or even high rates of return. The extent to which entrepreneurs take tax considerations into account is unclear; however, there is some reason to doubt that capital gains taxes are important in obtaining large amounts of venture capital, in part because much of this capital is supplied by those not subject to the capital gains tax (i.e., pension funds, foreign investors). (See CRS Report RL30040, *Capital Gains Taxes: Innovation and Growth*, by Jane G. Gravelle, for a further discussion of capital gains taxes and venture capital.) Moreover, there is no evidence that longer corporate stock holding periods lead to more investments in long-term assets, including R and D, a rationale for lowering rates for assets with longer holding periods. (Note that indexing, initially proposed, and then dropped, favors assets with shorter holding periods.)

A major complaint made by some about lower gains rates cut is that they primarily benefit very high income individuals. Capital gains are concentrated among higher income individuals because these individuals tend to own capital and because they are likely to own capital that generates capital gains. For example, the Joint Committee on Taxation indicated that for 2005, 88% of the benefit of lower rates to individuals with incomes over $200,000 and 95% would go to individuals with incomes over $100,000. Individuals with $200,000 of income account for about 3% of taxpayers and individuals with incomes over $100,000 account for less about 14%. The distributional effects of the capital gains relief for homes is somewhat less concentrated at the higher end (although lower income individuals are much less likely to own homes). The revisions may also enhance horizontal equity by treating taxpayer in different circumstances more evenly.

. Critics of lower capital gains taxes cite the contribution of preferential capital gains treatment to tax sheltering activities and complexity. For example, individuals may borrow (while deducting interest in full) to make investments that are eligible for lower capital gains tax rates, thereby earning high rates of return. These effects are, however, constrained in some cases (i.e., a passive loss restriction limits the deductions allowed in real estate ventures). Capital gains differentials complicate the tax law, especially as applied to depreciable assets where capital gains treatment can create incentives for churning assets unless recapture provisions are adopted. Indexing capital gains for inflation is more complicated than a simple exclusion, since different basis adjustments must apply to different vintages of assets and proper indexing of gains on depreciable property is especially difficult. Thus, foregoing indexing probably kept the tax law simpler.

In: Agricultural Finance and Credit
Editor: J. M. Bishoff, pp. 67-73

ISBN: 978-1-60456-072-5
© 2008 Nova Science Publishers, Inc.

Chapter 5

AGRICULTURAL CREDIT: INSTITUTIONS AND ISSUES*

Jim Monke

Analyst in Agricultural Policy Resources, Science, and Industry Division

ABSTRACT

The federal government has a long history of providing credit assistance to farmers by issuing direct loans and guarantees, and creating rural lending institutions. These institutions include the Farm Service Agency (FSA) of the U.S. Department of Agriculture (USDA), which makes or guarantees loans to farmers who cannot qualify at other lenders, and the Farm Credit System (FCS), which is a network of borrower-owned lending institutions operating as a government-sponsored enterprise.

The 110th Congress is expected to address agricultural credit through both appropriations and authorizations bills. Appropriators will consider funding for FSA's farm loan programs, and the agriculture committees may consider changes to FSA and FCS lending programs. The 2007 farm bill is expected to be the venue for many of the authorizing issues, although stand-alone legislation may be used for extensive reforms. This report will be updated.

BACKGROUND

The federal government has a long history of providing credit assistance to farmers. USDA's Farm Service Agency (FSA) issues direct loans and offers guarantees on loans made by commercial lenders. The direct and guaranteed loans are intended to assist farmer borrowers who do not qualify for regular commercial loans. Therefore, FSA is called a lender of last resort. The Farm Credit System (FCS), second only to commercial banks as a holder of farm debt, is chartered by the federal government as a cooperatively owned commercial lender to serve only agriculture-related borrowers. FCS makes loans to creditworthy farmers much like commercial banks, and is *not* a lender of last resort. Statutory authority for both the FSA and FCS lending programs is permanent, but omnibus farm bills, such as the expected

---

* Excerpted from CRS Report RS21977, dated March 8, 2007.

2007 farm bill, often make adjustments to the eligibility criteria and operations of the loan programs.

Other sources of credit for agriculture include commercial banks, life insurance companies, and individuals, merchants, and dealers. Figure 1 shows that commercial banks lend the largest portion of the farm sector's total debt (37%), followed by the Farm Credit System (30%), individuals and others (21%), and life insurance companies (5%). The Farm Service Agency provides 3% of the debt through direct loans, and guarantees another 4% of the market (through loans issued by commercial banks and FCS). Ranked by type of loan, the FCS has the largest share of real estate loans (38%), and commercial banks have the largest share of non-real estate loans (49%).

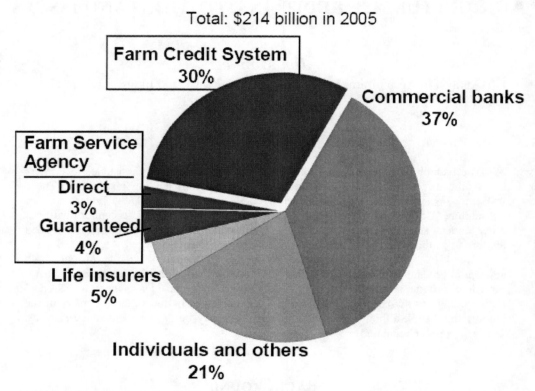

Source: CRS, using USDA-ERS and FSA data at [http://www.ers.usda.gov/Briefing/ FarmIncome / Data/Bs_t6.htm].

Figure 1. Market Shares of Farm Debt, by Lender.

Credit is an important input to agriculture, with all lenders holding about $214 billion in outstanding farm loans in 2005. Yet only about 66% of farmers have any debt (farm or nonfarm), and only 38% have farm debt. The types of farms holding the most debt include the larger commercial farms that produce most of the output, and medium-sized family farms.

Creditworthy farmers generally have adequate access to loans, mostly from the largest suppliers — commercial banks, FCS, and merchants and dealers. According to reports from lenders, credit conditions are good, and default rates have been trending lower to levels not seen since before the credit crisis of the 1980s. Overall, USDA data show that debt-to-asset ratios for the farm sector have been stable or slightly declining over the past decade, indicating that the sector is not highly leveraged with debt. Recent strength in farm income has given farmers more capacity to repay their loans or borrow new funds. Farm equity has been rising because increases in debt typically have been more than offset by larger gains in land values.

Nonetheless, despite the relatively strong farm economy in recent years, some farmers continue to experience financial stress due to individual circumstances, and may be unable to qualify for loans. Agriculture is also prone to business cycles that may pose financial difficulties. Thus, many interests in production agriculture continue to see some need for federal intervention in agricultural credit markets.

FARM LENDING INSTITUTIONS

Commercial Banks, Life Insurers, and Individuals

Together, commercial banks, life insurance companies, and individuals and others provide 63% of total farm debt without federal support or mandate. Commercial banks provide most of the loans to farmers through both small community banks and large multi-bank institutions.[1] Life insurance companies historically also have looked to farm real estate mortgages for diversification. Another important category of lenders is "individuals and others." This category consists of seller-financed and personal loans from private individuals, and the growing business segment of "captive financing" by equipment dealers and input suppliers (e.g., John Deere Credit and Pioneer Hi-Bred Financial Services).

Farm Credit System (FCS)[2]

Congress established the Farm Credit System in 1916 to provide a dependable and affordable source of credit to rural areas at a time when commercial lenders avoided farm loans. Operating as a government-sponsored enterprise, FCS is a network of borrower-owned lending institutions. It is not a government agency or guaranteed by the U.S. government. FCS is not a lender of last resort; it is a for-profit lender with a statutory mandate to serve agriculture. Funds are raised through the sale of FCS bonds and notes on Wall Street. Five large banks allocate these funds to 96 credit associations that, in turn, make loans to eligible creditworthy borrowers.

Statute and oversight by the agriculture committees determine the scope of FCS activity, and provide benefits such as tax exemptions. The system is regulated by the Farm Credit Administration (FCA). The program has permanent authority under the Farm Credit Act of 1971, as amended (12 U.S.C. 2001 *et seq.*). Major amendments generally have been enacted as stand-alone legislation, but Congress has used omnibus farm bills to make minor adjustments to the law.

FCS does not receive an annual appropriation, but is privately funded. Appropriators in recent years, however, have placed a limit on the size of the FCA's budget, which is funded by assessments on FCS institutions. For more background about FCS, see CRS Report RS21278, *Farm Credit System*, by Jim Monke.

USDA's Farm Service Agency (FSA)[3]

The USDA Farm Service Agency (FSA) is a lender of last resort because it makes direct loans to family-sized farms that are unable to obtain commercial credit.[4] FSA also guarantees timely payment of principal and interest on qualified loans made by commercial lenders such as banks and the Farm Credit System. The programs have permanent authority under the Consolidated Farm and Rural Development Act (CONACT, 7 U.S.C. 1921 *et seq.*). However, Congress uses omnibus farm bills to make changes to the terms, conditions, and eligibility requirements.

FSA makes farm ownership and operating loans to operators of family-sized farms. The maximum direct loans are $200,000 per borrower, while the maximum guaranteed loans are $852,000 per borrower (adjusted annually for inflation). Emergency loans are available for qualifying natural or other disasters. Some guaranteed loans have a subsidized (below-market) interest rate. To qualify for an FSA guaranteed or direct loan, farmers must demonstrate enough cash flow to make payments.

Since the 1980s, the emphasis within the FSA farm loan program has gradually shifted toward making relatively fewer direct loans and issuing more in guarantees. This lessens farmers' reliance on direct federal lending, and helps leverage federal dollars since guaranteed loans are cheaper to subsidize. In the late 1990s, about 30% of USDA farm loan authority was for direct loans. That ratio dropped to about 21% in FY2003, before rising again to about 25% in FY2004-FY2006.

Certain portions of the FSA farm loan program are reserved for beginning farmers and ranchers (7 U.S.C. 1994 (b)(2)). For direct loans, 70% of the amount for farm ownership loans and 35% of direct operating loans are reserved for beginning farmers for the first 11 months of the fiscal year (until September 1). For guaranteed loans, 25% is reserved for such farmers for ownership loans and 40% for farm operating loans for the first six months of the fiscal year (until April 1). Funds are also targeted to "socially disadvantaged" farmers based on race, gender, and ethnicity (7 U.S.C. 2003).[5]

As an example of the type of statutory changes made in a farm bill, Title V of the 2002 farm bill (P.L. 107-171) authorized funding levels for FSA loans for FY2003-FY2007 and expanded access to loans for beginning farmers. The 2002 law also increased the percentage that USDA may lend for real estate loan down-payments and extended the duration of eligible loans. It created a pilot program to guarantee seller-financed land contracts, available to five contracts per year in each eligible state (originally implemented in Indiana, Iowa, North Dakota, Oregon, Pennsylvania, and Wisconsin; in 2005, the program expanded to include California, Minnesota, and Nebraska).

Authorizations and Appropriations for Farm Loans

The 2002 farm bill authorized a maximum loan authority of $3.796 billion for direct and guaranteed loans for each of fiscal years 2003-2007 (7 U.S.C. 1994(b)(1)). Also, the law specified how this would be divided between direct and guaranteed loans, and within each of these categories how much could be used for ownership loans versus operating loans. The farm bill further instructed that not more than $750 million of the guaranteed operating loan amount may be used for the interest assistance (subsidized) guaranteed loan program (7 U.S.C. 1999), which reduces the interest rate on the loan by 4%.

Although the agriculture committee authorizes the farm bill with the multi-year "loan authority," the appropriations committee controls the annual discretionary appropriation to FSA to cover the actual cost of making loans (the "loan subsidy"). This loan subsidy is directly related to any interest rate subsidy provided by the government, as well as a projection of anticipated loan losses. The actual amount of lending that can be made (the appropriated loan authority) is several times larger than the appropriated loan subsidy.

For FY2007, the farm loan program is unchanged from FY2006 under the year-long continuing resolution (P.L. 110-5). $150 million in total loan subsidy is supporting the $3.52 billion in loan authority. This results in an effective "multiplier" of 23 ($23 dollars of loan authority for each $1 of loan subsidy). Guaranteed loans have higher multipliers than direct loans, and farm ownership loans have higher multipliers than operating loans. The highest multiplier in FY2006 is 208, for guaranteed farm ownership loans. The lowest is eight, for subsidized guaranteed operating loans, which have a 4% interest rate subsidy. Appropriations for the salaries and expenses of FSA personnel administering the loan program are $309 million in FY2007.

For FY2008, the Administration requests $3.37 billion in loan authority (-4.3%) to be supported by $152 million of loan subsidy (+1.7%). Guaranteed loan levels would decline less than 1% overall, although subsidized operating loans would decrease 8%. Greater reductions impact the direct farm loan program, which would decline 12%, including a decrease of 18% for direct ownership loans and 6% for direct farm operating loans. Despite the reduction in direct loan authority, subsidies for the direct loan programs would rise by over 10%. Administrative expenses would increase by 3.3%. As in recent years, nothing is requested for emergency loans due to carryover funds.

POLICY ISSUES FOR CONGRESS

Farm Service Agency

Authority for the size of FSA's farm loan program is specified in the 2002 farm bill and expires at the end of FY2007. The 2007 farm bill is seen as a vehicle to set new loan authorization levels for FSA, although actual funding would continue to be set by annual appropriations acts.

Some have expressed a desire to increase the $200,000 limit per farmer on direct farm ownership and operating loans.[6] These limits were set in 1984 for direct farm ownership loans, and in 1986 for direct operating loans, and have not kept pace with inflation. (Limits for guaranteed loans were raised in 1998 and indexed for inflation.)

Another potential issue is the "term limits" set in statute for farmer eligibility. Currently farmers are limited to receiving direct operating loan eligibility for seven years, and guaranteed operating loans for 15 years (7 U.S.C. 1949). A provision in the 2002 farm bill (Sec. 5102 of P.L. 107-171) suspended application of the 15-year limit through the end of 2006, and P.L. 109-467 extends the suspension provision until September 30, 2007. An

increasing number of farmers are reaching their term limits, and may face financial collapse if
they are not able to "graduate" to commercial credit. Term limits are intended to prevent
chronically inefficient farms from continuing to receive federally subsidized credit, but the
political and social consequences of letting these family farms fail are sometimes unpleasant.
Thus, there will be pressures to again extend the eligibility allowance or revisit the purpose of
the term limits requirement.

Farm Credit System

In recent years, FCS has expanded its lending, to a limited degree, beyond traditional
farm loans and into more rural housing and non-farm businesses. FCS also generally desires
to update the Farm Credit Act of 1971, which last was amended comprehensively in 1987. In
early 2006, FCS released a report titled *Horizons*, which highlights perceived needs for
greater lending authority to serve a changing rural America.[7] Some see Horizons as a
precursor to legislative action to expand lending authorities, possibly in the 2007 farm bill, or
to more regulatory changes expanding the allowed scope of lending.[8]

The scope of lending authority could grow under an October 2006 proposed rule to
expand eligibility for farm processing and marketing loans (71 FR 60678, October 16, 2006).
The intent appears to be to allow financing for larger value-added farm processing firms that
are being built with more outside capital and involvement than in previous decades.
Opponents fear that the regulation could allow more non-agriculture financing.

Selected FCS institutions also have begun investing in "agricultural and rural community
bonds" as a pilot project, with the approval of FCA. The bonds, issued by private or public
enterprises, are assets to the FCS institution with structured payment terms. The bonds
effectively result in loans to businesses and communities, some of which may not otherwise
qualify for FCS loans. For the FCS institution, the bonds are treated as an investment and thus
not subject to loan eligibility regulations.

Commercial banks oppose expanding FCS lending authority, saying that commercial
credit in rural areas is not constrained and that FCS's government-sponsored enterprise (GSE)
status provides an unfair competitive advantage. Commercial banks assert that, with financial
deregulation and integration, there is no credit shortage for agriculture and the federal benefits
for FCS are no longer necessary. FCS counters this by asserting its statutory mandate to serve
agriculture (and by extension, rural areas) through good times and bad, unlike commercial
lenders without such a mandate.

The controversy over GSE status and lending authority was highlighted in 2004 when a
private bank, Netherlands-based Rabobank, tried to purchase an FCS association. The board
of directors of Omaha-based Farm Credit Services of America (FCSA) initially voted for the
sale, indicating to some that FCS may no longer need government sponsorship. A general
outcry led FCSA to withdraw from the deal.[9] Commercial bankers say that institutions
should be allowed to leave FCS if they want more lending authorities. In 2004, FCS asked
Congress to eliminate the provision allowing institutions to leave the system (12 U.S.C.
2279d). It is not clear whether Congress, in 1987, intended the provision to be used by outside
companies to purchase parts of FCS. In 2006, the Farm Credit Administration amended the
rules governing how an FCS institution may terminate its charter (71 FR 44409, August 4,

2006). The changes allow more time for FCA to review the request, more communication, and more shareholder involvement.

REFERENCES

[1] Commercial bank issues are summarized by the American Bankers Association at [http://www. aba.com/Industry+Issues/issues_ag_menu.htm] and the Independent Community Bankers of America at [http://www.icba.org].

[2] Farm Credit System institutions are described at [http://www.fca.gov/FCS-Institutions.htm].

[3] USDA Farm Service Agency loan programs are described at [http://www. fsa.usda.gov/dafl].

[4] Historically, the USDA's lending agency was the Farmers' Home Administration (FmHA), created in 1945. A reorganization in 1995 moved the farm lending programs into FSA.

[5] Further background on FSA programs and delivery mechanisms are available in a USDA report to Congress, "Evaluating the Relative Cost Effectiveness of the Farm Service Agency's Farm Loan Programs," by Charles Dodson and Steven Koenig, at [http://www.fsa.usda.gov/ Internet/FSA_File/farm_loan_study_august_06.pdf]

[6] Glenn Keppy (Associate Administrator, USDA-FSA), testimony before Senate Agriculture Committee hearing, "Review USDA Farm Loan Programs," June 13, 2006, at [http://agriculture.senate.gov/Hearings/hearings.cfm?hearingId=1940].

[7] The *Horizons* report is available at [http://www.fchorizons.com].

[8] Bert Ely, "The Farm Credit System: Lending Anywhere but on the Farm," at [http://www.aba.com/NR/rdonlyres/E1577452-246C-11D5-AB7C-00508B95258D/45256/Horizons 2006ELYFINAL.pdf].

[9] For further background, see CRS Report RS21919, Farm Credit Services of America Ends Attempt to Leave the Farm Credit System, by Jim Monke.

In: Agricultural Finance and Credit
Editor: J. M. Bishoff, pp. 75-81

ISBN: 978-1-60456-072-5
© 2008 Nova Science Publishers, Inc.

Chapter 6

FARM CREDIT SERVICES OF AMERICA ENDS ATTEMPT TO LEAVE THE FARM CREDIT SYSTEM*

Jim Monke

Analyst in Agricultural Policy Resources, Science, and Industry Division

ABSTRACT

In an unprecedented move, an institution of the Farm Credit System (FCS) — a government-sponsored enterprise — initiated procedures on July 30, 2004, to leave the FCS and be purchased by a private company. But after much controversy, including congressional hearings, the board of directors of Farm Credit Services of America (FCSA) voted on October 19, 2004, to terminate its agreement with Rabobank before seeking approval from the Farm Credit Administration, the System's federal regulator.

FCSA is the FCS lending association serving Iowa, Nebraska, South Dakota, and Wyoming. Rabobank is a private Dutch banking company with extensive experience in agriculture and a growing global network. Under the plan, the loans, facilities, and employees of FCSA would have become part of Rabobank, and new FCS charters would have been issued to reestablish a System presence in the four-state region.

The option to leave the System is allowed by statute under the Farm Credit Act of 1971, as amended, but has been exercised only once, and did not involve an outside purchaser. Although Congress had no direct statutory role in the approval process, the House held hearings on the implications of the deal, and Senators Daschle and Johnson introduced S. 2851 to require public hearings and a longer approval process. This report will not be updated.

BACKGROUND ON THE FARM CREDIT SYSTEM

The Farm Credit System (FCS or System) is a national network of cooperatively owned lending institutions that provide credit and other services to farmers and ranchers. The FCS is a federally chartered institution, created in 1916 by Congress in the Federal Farm Loan Act. It

---

* Excerpted from CRS Report RS21919, dated October 22, 2004.

has a statutory mandate to serve agriculture as a permanent, reliable source of credit. Current statutory authority is in the Farm Credit Act of 1971, as amended. The most comprehensive recent changes were enacted in the Agricultural Credit Act of 1987 (P.L. 100-233). Federal oversight by the House and Senate Agriculture Committees in conjunction with regulations and examinations by the Farm Credit Administration (FCA) are designed to provide for the safety and soundness of System institutions. As a government-sponsored enterprise (GSE), the System has been given by Congress certain exemptions from taxation, and other benefits that presumably allow it to overcome barriers that might prevent purely private lenders from serving agriculture in the manner Congress envisioned. Unlike the housing GSEs, which are secondary markets, the FCS is a direct lender.

The System is a composed of four regional Farm Credit Banks (FCBs) and one Agricultural Credit Bank (ACB), each of which has chartered territory for serving farmers nationwide.[1] Funds from the sale of bonds flow through these five banks to 97 FCS lending associations, the second-largest of which in terms of assets is Farm Credit Services of America. FCS lending associations are cooperatives governed by directors elected from the borrowers who are also cooperative stockholders. They lend to farmers either directly or through their subsidiaries. For more information on the structure of the Farm Credit System, see CRS Report RS21278, *Farm Credit System*.

THE ABILITY TO LEAVE THE FARM CREDIT SYSTEM

Section 416 of the Agricultural Credit Act of 1987 (P.L. 100-233) amended the Farm Credit Act of 1971 to allow institutions to leave the Farm Credit System. These provisions originated in the Senate bill and were adopted by the conference committee (H.Rept. 100-490). The statute (12 U.S.C. 2279d) is implemented through detailed FCA regulations (12 C.F.R. 611.1200-1290) that specify the types of information that must be provided to FCA and the institution's shareholders throughout the termination process. By law, FCA must approve the plan before shareholders can vote to leave the System.

The main requirements of the termination procedure are as follows:

- Commencement Resolution. The association notifies FCA and stockholders of the plan to terminate and its effect on stockholders.
- Plan of Termination. The association submits a detailed plan to FCA including a proposed stockholder information statement, evidence of a new charter to be granted if FCS status is revoked, and an estimate of the exit fee. The exit fee is capital exceeding 6% of the association's assets.
- FCA Approval or Disapproval. If FCA disapproves, it must explain. One reason mentioned in the regulations is an "adverse effect on the ability of remaining System institutions to fulfill their statutory purpose."
- Stockholder vote. If FCA approves the plan, a majority of stockholders in the association who vote must approve the plan.
- Reconsideration petition. If the plan is approved by stockholders, a petition by 15% of stockholders may force a second and binding vote.

- Termination. If approved by stockholders, the association pays its debts and deposits the exit fee in escrow. FCA revokes the charter.
- Post-termination. FCA determines the exact exit fee.

The timeline for the above steps requires at least seven months. From the date the resolution is submitted, the association must wait at least 30 days to submit the termination plan. Once submitted, FCA has 60 days to consider the plan. If FCA approves, stockholders have 30 days to review the information statement before voting. If a majority approve, a 35-day period is allowed for a petition to re-vote. Termination can occur no sooner than 90 days after stockholder approval. Pending a termination, FCA would issue new lending charters so that the System could maintain a presence in the affected region. In this case, FCSA's plan did not proceed beyond the commencement resolution. After releasing its resolution on July 30, 2004, FCSA never submitted its termination plan before the board canceled the agreement on October 19, 2004.

The exit fee is a payment required in statute by the Farm Credit Act. The exit fee serves to reimburse the System for the capital earned from the benefits of being in the System, and is defined as capital exceeding 6% of assets over a multiyear period. FCA may review the association's records and make adjustments in calculating the final exit fee. This prevents an association from manipulating its capital to reduce the exit fee.

FCSA's attempt to leave the Farm Credit System was unprecedented in two ways: size, and purchase by an outside entity. Only one System institution has used the termination provisions. In 1991, the California Livestock Production Credit Association ($14 million in loans) became Stockmans Bank after becoming dissatisfied about making payments to prop up failing System institutions. Congress approved that termination in the 1990 farm bill and waived some fees (P.L. 101-624, Sec. 1838). It is not clear that Congress intended for a System institution to be purchased by an outside company.

THE OFFER

On October 19, 2004, the board of directors of Farm Credit Services of America (FCSA) voted to terminate its July 30, 2004, agreement with Rabobank to be purchased for $600 million payable to stockholders and a projected $800 million "exit fee" payable to the Farm Credit System Insurance Corporation. The board also voted on October 19 to reject a merger within FCS, and to initiate a patronage payment plan for its borrowers.

The offer generated significant controversy and congressional hearings over the financial terms and future of FCS. Rabobank had increased the stock offer to $750 million, following a $650 million offer from a neighboring FCS institution (see next section).

FCSA would have given up the benefits of membership in the Farm Credit System, including the tax exemption on its real estate loan portfolio and access to System funds. Shareholders would have owed capital gains taxes on the stock payment, and FCSA may have owed taxes on the exit fee attributed to the tax-exempt real estate portfolio.

Background on FCSA

As the second-largest of the System's 97 lending associations, FCSA is headquartered in Omaha, Nebraska, and has 43 offices and 51,000 shareholders. In 2003, its $7.3 billion loan portfolio was distributed geographically with 42% in Iowa, 39% in Nebraska, 16% in South Dakota, and 3% in Wyoming.[2]

Within the Farm Credit System, FCSA is one of the 18 lending associations in the AgriBank Farm Credit district, one of the System's five large regional banks. In March 2004, FCSA represented 6.3% of total combined System assets of $120.5 billion. In terms of loans to customers, FCSA held 8% of the System's $91 billion loan portfolio, and about 25% of the AgriBank district's loan portfolio.[3]

Historically, FCSA's four-state territory was the Omaha district, one of the twelve original Farm Credit districts. In 1994, the Farm Credit Bank of Omaha merged with the Farm Credit Bank of Spokane to become AgAmerica Farm Credit Bank, and the Omaha district consolidated into FCSA. On January 1, 2003, AgAmerica dissolved into two parts, and Farm Credit Services of America became part of the AgriBank district.

Background on Rabobank

Rabobank is a private Dutch banking cooperative with a long history of agricultural lending in the Netherlands. Rabobank has $500 billion in assets with operations in 35 countries. Rabobank has a 25-year history in the United States, generally financing larger agribusinesses and cooperatives. In recent years, Rabobank has moved into farm-level lending in the U.S. with the purchases of Valley Independent Bank (California) in 2002, Lend Lease Agri-Business (St. Louis) for $45 million in 2003, and Ag Services of America (Cedar Falls, Iowa) for $47 million in 2003.

ALTERNATIVE OFFER WITHIN THE SYSTEM

Before the FCSA board accepted Rabobank's offer on July 30, AgStar Financial Services, an FCS association in Minnesota, made an undisclosed offer to merge with FCSA and keep it in the System. On August 18, 2004, AgStar submitted another offer to purchase FCSA for $650 million, $50 million more than Rabobank's initial offer. AgStar's territory is adjacent to FCSA, and includes the southern and eastern halves of Minnesota, and the northwestern portion of Wisconsin. With 12,000 stockholders and $2.4 billion in loans, AgStar is smaller than FCSA.

The AgStar offer was meant to be competitive with Rabobank in terms of the stockholder payment. But rather than a buyout, the AgStar offer was a merger of two System institutions. The $650 million payment would be more of a patronage or dividend distribution rather than a stock buyout, and shareholders would continue to be owners in the merged association. AgStar stated that it would make patronage payments to FCSA shareholders, something that current FCSA management had not done. No exit fee would be required with an AgStar merger, allowing $800 million in capital to remain in the association, rather than being transferred to the nationwide FCS Insurance Corporation.[4]

IMPLICATIONS FOR THE FUTURE OF THE FARM CREDIT SYSTEM

Regional Implications

With FCSA terminating its bid to exit the System, service to FCS customers in the region should continue uninterrupted through FCSA.

However, if FCSA had been purchased by Rabobank, FCA would have issued new charters and the Farm Credit System would have needed to rebuild a physical infrastructure of offices and employees, as well as its portfolio of loans and customers. Thus, even though issuing new charters could have maintained a System presence in the region, the magnitude of that presence could have been significantly smaller for some time, depending on employee and customer loyalties.

National Implications

Although the Rabobank agreement directly affected service to only a fraction of the System, the implications for the entire System have been greater. FCSA's acceptance of Rabobank's offer has given opponents of the System additional reasons to question the rationale supporting the System's existence.

When Congress passed the Federal Farm Loan Act in 1916, credit was frequently unavailable or unaffordable in rural areas. Many lenders avoided agricultural loans due to the inherent risks. Thus, Congress created the Farm Credit System and provided certain financial benefits to assure a permanent, reliable source of credit to American agriculture.

For more than a decade, credit has been available to most farmers from a variety of sources, including commercial banks, life insurance companies, farm input suppliers, the U.S. Department of Agriculture (USDA), and the Farm Credit System. Furthermore, the reliability of government commodity payment programs has given agricultural lenders extra assurance that most farm borrowers will be able to repay their loans.

These factors have caused some observers to question whether the same need exists today for FCS as in the early part of the 20th century. Such critics of FCS say preferential treatment is not warranted since agriculture no longer faces a credit constraint and other industries do not receive such treatment.[5]

Thus, the attempt by Farm Credit Services of America to voluntarily become private is being seen by some as an indicator that the System may no longer need its government sponsorship. The American Bankers Association has asserted for many years that the FCS no longer warrants its GSE status and is now citing this buyout offer as further evidence for that position.[6] Such commercial lenders are among the only groups that did not express opposition to the sale, provided that taxpayer interests were adequately addressed.

The System counters arguments over its GSE status by asserting its statutory mandate to serve agriculture through both good times and bad. The Farm Credit Council, the System's lobbying arm, was opposed to the Rabobank purchase, and contended that farmer borrowers would be better served under the status quo or under the AgStar merger.[7] Most farm groups expressed concern over the proposed sale to a private, and foreign, company.

ISSUES FOR CONGRESS

Congress had no statutory role in the termination process. However, some Members of Congress took an interest, especially given concerns by some farmers over the future of the Farm Credit System generally and more specifically in the four-state region. Some Members' offices in states outside the affected region received constituent mail about the future of their loans, and whether the rest of the System "is for sale."

A House Agriculture Subcommittee on Conservation, Credit, Rural Development, and Research held a hearing on the issue on September 29, 2004. The Farm Credit System asked Congress to remove the statutory language allowing institutions to leave the System. Commercial bankers asked for better access to FCS funding, and testified that institutions should be allowed to exit the System if they want more lending authorities than allowed under the Farm Credit Act.

In the Senate, Senators Daschle and Johnson of South Dakota introduced S. 2851 to require FCA to hold public hearings on the implications of the proposed purchase, and to increase FCA's review period from a maximum 60-day period to a minimum six-month period.

In conclusion, the option for FCS institutions to terminate their status in the System is allowed in statute and regulation. Despite aborting its attempt to leave the System, Farm Credit Services of America's decision to be bought by a private firm may affect the agricultural and lending industry's view of the Farm Credit System into the future. The agreement served to highlight certain provisions in the Farm Credit Act that both proponents and opponents of the System say need attention. Future efforts to address these issues may be affected by these recent events.

REFERENCES

[1]   For a directory of institutions in the Farm Credit System, and a map of the five regional banks, see the Farm Credit Administration website at [http://www.fca.gov/apps/ instit.nsf].

[2]   FCSA, *2003 Annual Report,* [http://www.fcsamerica.com/company/AR99FCSA.pdf].

[3]   AgriBank, FCB, *2003 Annual Report*, [http://www.agribank.com/Docs/03-Annual-bank.pdf].

[4]   For more analysis, see "FCS of America's Organizational Choices" by Peter Barry (University of Illinois) [http://www.farmdoc.uiuc.edu/finance/publications/ FCSA %20Rabobank%20Agstar% 20choices.pdf] and "Understanding the Proposed Sale of Farm Credit Services of America" by Neil Harl, et al. (Iowa State University) [http://www.econ.iastate.edu/ rabobankbuyout].

[5]   For example, see Bert Ely, "The Farm Credit System: Reinvented and Mission-Challenged," November 2002, at [http://www.aba.com/Industry+Issues/ issues_ag_menu.htm].

[6]   See American Bankers Association, press release, July 30, 2004, at [http://www.aba.com/Press+Room/073004statement.htm].

[7]   See Farm Credit Council, press release, July 30, 2004, at [http://www.fccouncil.com/press/ fcsofamerica.pdf].

In: Agricultural Finance and Credit
Editor: J. M. Bishoff, pp. 83-89

ISBN: 978-1-60456-072-5
© 2008 Nova Science Publishers, Inc.

Chapter 7

FARM CREDIT SYSTEM*

Jim Monke

Analyst in Agricultural Policy Resources, Science, and Industry Division

ABSTRACT

The Farm Credit System (FCS) is a nationwide financial cooperative lending to agricultural and aquatic producers, rural homeowners, and certain agriculture-related businesses and cooperatives. Established in 1916, this government-sponsored enterprise (GSE) has a statutory mandate to serve agriculture. It receives tax benefits, but no federal appropriations or guarantees. FCS is the only direct lender among the GSEs. Farmer Mac, a separate GSE but regulated under the umbrella of FCS, is a secondary market for farm loans. Federal oversight by the Farm Credit Administration (FCA) provides for the safety and soundness of FCS institutions.

Current issues and legislation affecting the FCS are discussed in CRS Report RS21977, *Agricultural Credit: Institutions and Issues*. This report will be updated.

WHAT IS THE FARM CREDIT SYSTEM?

An Agricultural Lender

The Farm Credit System (FCS) was created to provide a permanent, reliable source of credit to U.S. agriculture. When Congress enacted the Federal Farm Loan Act in 1916, credit was frequently unavailable or unaffordable in rural areas. Many lenders avoided such loans due to the inherent risks of agriculture. Statutory authority is in the Farm Credit Act of 1971, as amended (12 U.S.C. 2001 *et seq.*). The most comprehensive recent changes were enacted in the Agricultural Credit Act of 1987.

The FCS is authorized by statute to lend to farmers, ranchers, and harvesters of aquatic products. Loans may also be made to finance the processing and marketing activities of these

---

* Excerpted from CRS Report RS21278, dated January 25, 2007.

borrowers, for home ownership in rural areas, certain farm- or ranch-related businesses, and agricultural, aquatic, and public utility cooperatives.

The FCS is *not* a lender of last resort.[1] FCS is a commercial, for-profit lender. Borrowers must meet creditworthiness requirements similar to those of a commercial lender. FCS has "young, beginning, and small" (YBS) farmer lending programs, but does not have particular targets or numerical mandates for such loans.

The FCS holds about 31% of the farm sector's total debt (second to the 40% share of commercial banks) and has the largest share of farm real estate loans (38%). As of September 2006, FCS had $115 billion in loans outstanding, of which about 47% was in long-term agricultural real estate loans, 24% in short- and intermediate-term agricultural loans, 15% in loans to agribusinesses, 5% in energy loans, 3% in rural home loans, and 6% in communications, export financing, leases, and water and waste disposal loans.

A Government-Sponsored Enterprise (GSE)

As a GSE, FCS is a privately owned, federally chartered cooperative designed to provide credit nationwide. It is limited to serving agriculture and related businesses and homeowners in rural areas. Each GSE is given certain benefits such as implicit federal guarantees or tax exemptions, presumably to overcome barriers faced by purely private markets.[2] FCS is the only direct lender among the GSEs; other GSEs such as Fannie Mae are secondary markets. FCS is not a government agency and its debt instruments and loans are not explicitly guaranteed by the U.S. government.[3]

The tax benefits for FCS include an exemption from federal, state, municipal, and local taxation on the profits earned by the real estate side of FCS (12 U.S.C. 2098). Income earned by the non-real estate side of FCS is subject to taxation. The exemption originated in the 1916 act. Commercial bankers estimate that the annual value of these tax benefits amounted to at least $425 million in 2004.[4] For investors who buy FCS bonds on Wall Street, the interest earned is exempt from state, municipal, and local taxes. This makes FCS bonds more attractive to the investing public and helps assure a plentiful supply of funds for loans. Commercial bankers say that the tax benefits let FCS offer lower interest rates to borrowers, and thus give FCS an operating advantage since they compete in the same retail lending market.

A Cooperative Business Organization

FCS associations are owned by the borrowers who purchase stock, typically as part of their loan. FCS stockholders elect the boards of directors for banks and associations. Each has one vote, regardless of the loan size. Most directors are members, but federal law requires at least one from outside.

If an association is profitable, the directors may choose to retain the profits to increase lending capital, or distribute some of the net income through dividends or *patronage refunds*, which are proportional to the size of loan. Patronage refunds effectively reduce the cost of borrowing. Some associations tend to regularly pay patronage while others prefer to retain their earnings or charge lower interest rates.

A National System of Banks and Associations

FCS is composed of five regional banks that provide funds and support services to 95 smaller Agricultural Credit Associations (ACAs), Federal Land Credit Associations (FLCAs), and Production Credit Associations (PCAs). These associations, in turn, provide loans to eligible borrowers. The most common operating structure (due to favorable tax and regulatory rules) is a "parent ACA" with FLCA and PCA subsidiaries. There are 86 ACAs and nine standalone FLCAs.[5]

One of the regional banks, CoBank, has a nationwide charter to finance farmer-owned cooperatives and rural utilities. It finances agricultural exports and provides international services for farmer-owned cooperatives through three international offices.

Capitalized with Bonds and Stock, Not the U.S. Treasury

The Federal Farm Credit Banks Funding Corporation ([http://www.farmcredit-ffcb.com]) uses capital markets to sell FCS bonds and notes. These debts become the joint and several liability of all FCS banks. The funding corporation allocates capital to the banks, which provide funds to associations, which lend to borrowers. Profits from loans repay bondholders.

FCS also raises capital through two other methods. Borrowers are required to pay the lesser of $1,000 or 2% of the loan amount and become a cooperative stockholder. FCS also retains profits that are not returned as patronage to borrowers.

With the exception of seed money that was repaid by the 1950s and a temporary U.S. Treasury line of credit in the 1980s,[6] FCS operates without any direct federal money. FCS banks and associations do not take deposits like commercial banks, nor do they receive federal appropriations to fund their loan program.

TYPES OF LOANS AND BORROWERS

The FCS provides three types of loans: (1) operating loans for the short-term financing of consumables such as feed, seed, fertilizer, or fuel; (2) installment loans for intermediate-term financing of durables such as equipment or breeding livestock; and (3) real estate loans for long-term financing (up to 40 years) of land, buildings, and homes.

The FCS has a statutory mandate to serve agriculture, and certain agribusinesses and rural homeowners. Borrowers must meet certain eligibility requirements in addition to general creditworthiness. Eligible borrowers and their scope of their financing can be grouped into four categories.

- Full-time farmers. For individuals with over 50% of their assets and income from agriculture, FCS can lend for all agricultural, family, and non-agricultural needs (including vehicles, education, home improvements, and vacation expenses).
- Part-time farmers. For individuals who own farmland or produce agricultural products but earn less than 50% of their income from agriculture, FCS can lend for all agricultural and family needs. However, non-agricultural lending is limited.

- Farming-related businesses.[7] FCS can lend to businesses that *process* or *market* farm, ranch, or aquatic products if more than 50% of the business is owned by farmers who provide at least some of the "throughput." FCS also can lend to businesses that *provide services* to farmers and ranchers (but not aquatic producers), such as crop spraying and cotton ginning. The extent of financing is based on the amount of the business's farm-related income.
- Rural homeowners. FCS can lend for the purchase, construction, improvement, or refinancing of single-family dwellings in rural areas with populations of 2,500 or less.

CONSOLIDATION

The number of banks and associations has been declining for decades through mergers and reorganizations. This consolidation accelerated, however, in 1999 when the Farm Credit Administration (FCA), the system's regulator, approved the "parent ACA" structure and the Internal Revenue Service declared FLCA subsidiaries tax-exempt. In the mid-1940s, there were over 2,000 lending associations, nearly 900 in 1983, fewer than 400 by 1987, 200 in 1998, and only 95 in 2006. The system operated with 12 districts into the 1980s, 8 districts in 1998, and 5 regional banks (districts) since 2004.

Twenty years ago, the typical FCS association covered several counties and specialized in either land or farm production loans. Today, the typical FCS association covers a much larger region, delivers a wide range of farm and rural credit programs and services, and has an extensive loan portfolio. FCS benefits when consolidation creates more diversified portfolios. Customers may benefit if greater institutional efficiency is passed along through lower interest rates. However, consolidation may weaken the original cooperative concept of local borrower control and close many local offices at which farmers had established relationships.

CHARTER TERRITORIES

Each association within FCS has a specific "charter territory." If an association wants to lend outside its charter territory, it first must obtain approval from the other territory's association. For example, associations within U.S. AgBank's region (the southern Plains and West) can compete for loans, but associations in the AgFirst region (the East and Southeast) cannot. Charter territories help ensure that borrowers are served locally and maintain local control of the association. Charter territories and any changes must be approved by FCA.

In 2001, FCA proposed allowing national charters so that associations would not be restricted by geographical boundaries. The FCA board later dropped the idea after opponents raised concerns that national charters would weaken FCS's mission by pitting associations against each other for prime loans and reducing commitments to local areas.

FEDERAL REGULATION

Congressional Oversight

Congressional oversight of FCS is provided by the House and Senate Agriculture Committees. The most recent hearings on FCS include one in the House on June 2, 2004, concerning Farmer Mac, and another in the House on September 29, 2004, over the proposed sale of an FCS association.

Farm Credit Administration (FCA)

The FCA ([http://www.fca.gov]) is an independent agency and the federal regulator responsible for examining and ensuring the safety and soundness of all FCS institutions. Regulations are published in 12 C.F.R. 600 *et seq*. FCA's operating expenses are paid through assessments on FCS banks and associations. Even though FCA does not receive an appropriation from Congress, the annual agriculture appropriations act in recent years has put a limit on FCA's administrative expenses ($44.25 million in FY2006).

FCA is directed by a three-member board nominated by the President and confirmed by the Senate. Board members serve a six-year term and may not be reappointed after serving a full term or more than three years of a previous member's term. The President designates one member as chairman, who serves until the end of that member's term.[8]

OTHER FCS ENTITIES

The Farm Credit System has several other entities besides the previously discussed banks and associations that lend money, the Federal Farm Credit Banks Funding Corporation, which sells FCS bonds on Wall Street, and the Farm Credit Administration, which examines and regulates the system.

Federal Agricultural Mortgage Company (Farmer Mac)

Farmer Mac ([http://www.farmermac.com]) was established in the Agricultural Credit Act of 1987 to serve as a secondary market for agricultural loans — purchasing and pooling qualified loans, then selling them as securities to investors. Farmer Mac increases the capacity for agricultural lenders to make more loans; for example, if a lender makes a 30-year loan and sells it to Farmer Mac, the proceeds can be used to make another loan.

Although Farmer Mac is part of FCS and regulated by FCA, it has no liability for the debt of any other FCS institution, and the other FCS institutions have no liability for Farmer Mac debt. It is considered a separate GSE. Farmer Mac is organized as an investor-owned corporation, not a member-owned cooperative. Voting stock may be owned by commercial banks, insurance companies, other financial organizations, and FCS institutions. Nonvoting stock may be owned by any investor. The board of directors has 15 members: five elected from the FCS, five elected from commercial banks, and five appointed from the public at large.

Farmer Mac operates two programs: Farmer Mac I (loans not guaranteed by USDA) and Farmer Mac II (USDA-guaranteed loans).

- A majority of Farmer Mac I volume comes from the sale of "long-term standby purchase agreements" (LTSPC). Farmer Mac promises to purchase specific agricultural mortgages, thus guaranteeing the loans against default risk while the participating lender retains interest rate risk.
- Under Farmer Mac II, the company purchases the portion of individual loans that are guaranteed by USDA. On these purchases, Farmer Mac accepts the interest rate risk but carries no default risk.

Farmer Mac continues to hold most of the loans it purchases, a potentially more profitable activity for the company, but also more risky.

Farm Credit System Insurance Corporation

The Insurance Corporation ([http://www.fcsic.gov]) was established by statute in 1988 to ensure timely payment of principal and interest on FCS debt securities. The FCA board comprises its board of directors. Annual premiums are paid by each bank through an assessment based on loan volume until the secure base amount of 2% of total outstanding loans is reached.

Farm Credit Council

The Farm Credit Council ([http://www.fccouncil.com]) is the national trade association of FCS. FCC has offices in Washington, DC, and Denver, CO, and lobbies on behalf of FCS. FCC also provides support services.

REFERENCES

[1]   The USDA Farm Service Agency (FSA) is a lender of last resort for borrowers who are unable to get a loan from another lender. For more information about other farm lenders and current issues, see CRS Report RS21977, *Agricultural Credit: Institutions and Issues*, by Jim Monke.

[2]   There are five GSEs: Federal National Mortgage Association (Fannie Mae), Federal Home Loan Mortgage Corporation (Freddie Mac), Federal Home Loan Bank System, Federal Agricultural Mortgage Corporation (Farmer Mac), and FCS. For more on GSEs, see CRS Report RL30533, *The Quasi Government: Hybrid Organizations with Government and Private Characteristics*.

[3]   Because of the significant role of GSEs in the U.S. economy, most investors feel the federal government will not allow a GSE to fail. Thus, an implicit, albeit not statutory, guarantee exists.

[4]   Bert Ely, *Farm Credit Watch*, June 2005.

[5]   For a directory of institutions in the Farm Credit System, and a map of the five regional banks, see the Farm Credit Administration website at [http://www.fca.gov/apps/instit.nsf].

[6]   The Financial Assistance Corporation (FAC) borrowed $1.26 billion from Treasury following the farm financial crisis of the 1980s. On June 10, 2005, the last of these bonds was repaid on schedule, and interest was repaid to the U.S. Treasury. The FAC will be dissolved by June 2007.

[7]   A proposed rule by the Farm Credit Administration could expand the eligibility criteria for farm processing and marketing loans (71 FR 60678, October 16, 2006). For more background, see CRS Report RS21779, *Agricultural Credit: Institutions and Issues*, by Jim Monke.

[8]   FCA board members are Nancy C. Pellett (chairman since May 22, 2004), appointed in 2002 for a term that ends in May 2008; Dallas Tonsager, appointed in 2004 for a term that ends in May 2010; and Leland Strom, appointed in 2006 for a term that ends in October 2012.

In: Agricultural Finance and Credit
Editor: J. M. Bishoff, pp. 91-95

ISBN: 978-1-60456-072-5
© 2008 Nova Science Publishers, Inc.

Chapter 8

CHAPTER 12 OF THE U.S. BANKRUPTCY CODE: REORGANIZATION OF A FAMILY FARMER OR FISHERMAN*

Robin Jeweler
Legislative Attorney American Law Division

ABSTRACT

Chapter 12 of the U.S. Bankruptcy Code dealing with "family farmer" reorganization, temporarily extended 11 times since its original enactment, is made permanent by enactment of the Bankruptcy Abuse Prevention and Consumer Protection Act, P.L. 109-8. It is amended to include "family fisherman" as well. This report surveys the highlights of this chapter.

BACKGROUND

In 1986, Congress added chapter 12 entitled "Adjustments of Debts of a Family Farmer with Regular Annual Income" to the U.S. Bankruptcy Code.[1] It was modeled after chapter 13, which governs consumer reorganization. Chapter 12 was created to provide farmers with the opportunity to reorganize and thus to preserve their farms through a streamlined and expeditious bankruptcy process. Intended to respond to the downturn in the farm economy in the 1980s, it was considered "experimental."[2] Originally enacted as a temporary measure with a sunset date of October 1, 1993,[3] it was extended 11 times.[4] With enactment of the Bankruptcy Abuse Prevention and Consumer Protection Act (BAPCPA) it became a permanent chapter of the U.S. Bankruptcy Code.[5] The BAPCPA made additional changes to chapter 12, which are discussed below.

In the absence of chapter 12, insolvent farmers' bankruptcy options included reorganization under chapter 11 or 13, or liquidation under chapter 7.[6] Chapter 11 governs

---

* Excerpted from CRS Report RS20742, dated August 2, 2005.

business reorganization and is more expensive and procedurally cumbersome than chapter 12 or 13. Chapter 13 was designed for "individuals with a regular income" or wage earners. The nature of farming is such that many farmers did not realize income on a "regular" schedule comparable to that of wage earners. Also, many farms were not individually owned, but are operated as corporations or partnerships. And, when chapter 12 was enacted in 1986, chapter 13 had a jurisdictional debt limit of $350,000 in secured debt and $100,000 in unsecured debt. Farmers' indebtedness often exceeded that limit, so chapter 12 provided a jurisdictional debt limit of $1,500,000.

The BAPCPA made major revisions to chapter 13, which make it even more unsuitable for farmers than it was previously. A new requirement for chapter 13 debtors is that disposable income reasonably necessary for living expenses (retained by a debtor and deducted from the amount allocated to repayment of creditors) be calculated according to national and local living standards calculated by the Internal Revenue Service (IRS). This requirement alone would appear to make chapter 13 wholly unsuited for family farmers and fishermen.

PROCEDURAL OVERVIEW

The goal of chapter 12, indeed of all the operative bankruptcy reorganization chapters, is to offer a debtor a means of financial rehabilitation outside of liquidation. Historically, the *quid pro quo* for reorganization is the debtor's obligation to commit postbankruptcy income to prebankruptcy indebtedness.[7]

Definition of "Family Farmer" and "Fisherman"

Previously applying only to "family farmers," the BAPCPA amended the chapter to include fishermen. For chapter 12 purposes, a family farmer includes an individual and spouse, or a family-owned partnership or corporation, with debts of less than $3,237,000, 50% of which arises from the farming operation. The debtor must derive at least 50% of gross annual income from farming.[8]

A "family fisherman" is an individual and spouse, or a family-owned partnership or corporation, engaged in a commercial fishing operation whose debts are less than $1,500,000, 80% of which arises from the fishing operation. The debtor must derive at least 50% of gross annual income from fishing.[9]

Expedited Time Frame

A chapter 11 debtor has at least 120 days (four months) after filing to submit a proposed reorganization plan. A chapter 12 debtor must submit a proposed plan within 90 days of filing,[10] and the court must hold a confirmation hearing within 45 days.[11] Only the debtor may propose the reorganization plan, which must be completed within a specified three to five-year time frame. The debtor's plan must meet the statutory requirements for chapter 12, but, unlike chapter 11, creditor committees are not appointed, and the plan is not voted on by creditors. The debtor receives a discharge of indebtedness upon completion of all payments under the plan.

Appointment of a Standing Trustee

A trustee is rarely appointed under chapter 11. Although chapter 12 requires the participation of a standing trustee, the trustee's duties are administrative and the debtor retains possession and control of the farm or fishing operation throughout the reorganization.

Benefits of Chapter 12

As noted above, chapter 12 borrowed many debtor-friendly features from chapter 13 (prior to the BAPCPA amendments) and adapts some provisions specifically to the needs of chapter 12 debtors. Among these provisions are:

- *Prebankruptcy credit counseling.* Although individual chapter 12 debtors must undergo credit counseling from an approved nonprofit agency within 180 days prior to filing, their ability to receive a discharge is not conditioned upon completion of an instruction course on personal financial management (as required of chapter 13 debtors);[12]
- *Contents of the plan: nonpriority treatment of certain priority claims.* The plan may treat certain claims owed to the government that might otherwise be considered "priority" claims under § 507 as nonpriority, namely, claims owed to a governmental unit that arise as a result of the sale or transfer of any farm asset used in the farming operation. Likewise, priority claims for domestic support that have been assigned to a governmental unit may be paid in less than full amount so long as the plan provides that all of the debtor's projected disposable income for five years will be applied to payments under the plan. Specified interest payments on unsecured claims that are nondischargeable may also be minimized;[13]
- *Liberalized "cramdown."* Cramdown is a term used in bankruptcy law to refer to the procedure which enables a debtor to modify a creditor's claim over the creditor's objection. Chapter 12 debtors are not bound to "the absolute priority rule" which applies in chapter 11. Hence, secured creditors of a chapter 12 debtor, despite objections they may harbor, have less ability to influence or reject the reorganization plan than chapter 11 creditors. Dissenting creditors may object to the debtor's reorganization plan. The objection triggers the Code's requirement that all of the debtor's projected disposable income be applied to plan payments. Creditors must receive at least as much payout on their claims in chapter 12 as they would if the debtor were liquidated.[14]
- *Bifurcation of liens or "lien stripping."[15]* A family farmer is permitted to reduce — or "strip" — the value of a lien on secured property.[16] Hence, in situations where there is deflation in the value of the farm property, the debtor may reduce the amount owed to the secured lender and discharge the unsecured portion. Lien

stripping is not generally permitted in chapter 7.[17] Although lien stripping is permitted under chapter 11, creditors whose security interests are bifurcated into secured and unsecured claims have greater power to shape and/or veto the debtor's reorganization plan. Chapter 12 also permits repayment of reformed secured claims over a period of time exceeding the plan period. Because chapter 12 provides the debtor with expanded authority to renegotiate with lenders, its influence may extend to informal renegotiations outside of bankruptcy as well.

- *Adequate protection and tax requirements.* Any bankruptcy debtor who retains collateral must provide the creditor "adequate protection" for the value of the collateral subject to the debtor's use.[18] Chapter 12 has its own standards for awarding "adequate protection"[19] and a provision governing "special tax provisions."[20]

Although chapter 12 was enacted as a temporary measure, it was continuously extended prior to becoming a permanent chapter in the Code. It has received significant support in the Congress. The chapter, however, is not without its critics. They question whether the benefits of chapter 12 outweigh its costs; whether it encourages economic inefficiency in the farm sector; and whether the ability of farmers to write off secured debt has a positive impact on agricultural lending.[21]

REFERENCES

[1]   P.L. 99-554 (Oct. 27, 1986).
[2]   H.Rept. 103-32, 103d Cong., 1st Sess. 3 (1993).
[3]   P.L. 99-554, §302(f).
[4]   P.L. 103-65 (extension through Sept. 30 1998); P.L. 105-277 (extension through April 1, 1999); P.L. 106-5 (extension through October 1, 1999); P.L. 106-70 (extension through July 1, 2000); P.L. 107-8 (extension through June 1, 2001); P.L. 107-17 (extension through October 1, 2001); P.L. 107-170 (extension through June 1, 2002); P.L. 107-171 (extension through January 1, 2003); P.L. 107-377 (extension through July 1, 2003); P.L. 108-73 (extension through Jan. 1, 2004); and P.L. 108-369 (extension through July 1, 2005). With the exception of P.L. 107-377 and P.L. 107-171, which applied the extension prospectively, other extensions applied retroactively.
[5]   P.L. 109-8, § 1001 (2005).
[6]   Farmers may not be forced into bankruptcy involuntarily. 11 U.S.C. § 303.
[7]   Under chapter 7, prebankruptcy assets are liquidated to pay off prebankruptcy debts. A chapter 7 liquidation does not generally commit the debtor's postbankruptcy assets and/or income, nor will it discharge postbankruptcy debts.
[8]   11 U.S.C. § 101(18).
[9]   11 U.S.C. § 101(19).
[10]   11 U.S.C. § 1221.
[11]   11 U.S.C. § 1224.
[12]   11 U.S.C. § 109.
[13]   11 U.S.C. § 1222.

[14] 11 U.S.C. § 1225.

[15] When a secured creditor holds a claim in which the value of the collateral is less than the contractual amount owed the creditor, the Bankruptcy Code splits or "bifurcates" it. It is treated as a secured claim up the value of the collateral, and an unsecured claim for the amount which constitutes the discrepancy between the collateral's value and the contractual amount of the claim. 11 U.S.C. § 506. In other words, the claim of a secured creditor is legally protected up to the value of the collateral which secures the claim.

[16] Harmon v. United States Through Farmers Home Admin., 101 F.3d 574 (8th Cir. 1996).

[17] Dewsnup v. Timm, 502 U.S. 410 (1992).

[18] 11 U.S.C. § 361.

[19] 11 U.S.C. § 1205.

[20] 11 U.S.C. § 1231.

[21] Jerome Stam, "Do Farmers Need a Separate Chapter in the Bankruptcy Code," Information Bulletin No. 724-09, Economic Research Service, USDA (October 1997).

In: Agricultural Finance and Credit
Editor: J. M. Bishoff, pp. 97-100

ISBN: 978-1-60456-072-5
© 2008 Nova Science Publishers, Inc.

Chapter 9

ECONOMIC ISSUES SURROUNDING THE ESTATE AND GIFT TAX: A BRIEF SUMMARY*

Jane G. Gravelle

Senior Specialist in Economic Policy Government and Finance Division

ABSTRACT

Supporters of the estate and gift tax argue that it provides progressivity in the federal tax system, provides a backstop to the individual income tax, and appropriately targets assets that are bestowed on heirs rather than assets earned through their hard work and effort. Progressivity, however, can be obtained through the income tax and the estate and gift tax is an imperfect backstop to the income tax. Critics argue that the tax discourages savings, harms small businesses and farms, taxes resources already subject to income taxes, and adds to the complexity of the tax system. Critics also suggest death is an inappropriate time to impose a tax. The effect on savings, however, is uncertain, most farms and small businesses do not pay the tax, and complexity could be reduced through reform of the tax. This report will be updated as legislative developments warrant.

The estate and gift tax has been the subject of legislative interest for several years, with increases in the exemption enacted in 1997. Proposals to reduce or eliminate the tax were adopted in the 106th Congress, but were vetoed by the President. President Bush had also proposed eliminating the tax and the Ways and Means Committee reported out a bill, H.R. 6, that would phase out the tax. Similar provisions were included in the Senate bill and the final tax cut bill, H.R. 1836, was signed by the President on June 7, 2001, although this legislation retained a gift tax with a large exemption.[1] The entire bill is to sunset after 2010, but there are proposals to make the change permanent, including H.R. 8 which passed the House on April 13, 2005. Further consideration to making the tax change permanent was originally scheduled for the fall of 2005, but was delayed because Congress was considering legislation relating to Hurricane Katrina.

---

* Excerpted from CRS Report RS20609, dated January 23, 2006.

The estate and gift tax is a comparatively small source of federal tax revenue, accounting for 1.2% of federal receipts, an amount which has declined following the reduction in rates and the increase in exemptions enacted in 2001. Estates and gifts to spouses are exempt from tax as are gifts to charity. The first $2 million of the net estate value is exempt from tax. Gift taxes have a $1 million effective exclusion in addition to an annual gift exclusion of $11,000 per donee. Taxable estates are subject to a 46% rate. The exemption from tax is scheduled to rise and the tax rate to fall with the estate tax (but not the gift tax) permanently eliminated in 2010, when a tax on capital gains in excess of an exempt amount will be imposed. The 2001 tax cuts, absent legislation, will sunset after 2010 and the rates and exemptions return to the values prior to the 2001 tax changes. The House recently passed legislation to make the repeal permanent.

ARGUMENTS FOR THE ESTATE AND GIFT TAX

Perhaps the principal argument in support of an estate and gift tax is its contribution to progressivity in the income tax system. The estate tax is the most progressive of any of the federal taxes. According to the latest data from the Internal Revenue Service (2004), out of the approximately 2.4 million deaths per year, only 1.3 % of decedents' estates paid the estate tax; based on the distribution of gross estate, that share is probably close to 0.5% for 2006 and will drop to about 0.2% in 2009. These numbers can be contrasted with the income tax where about 70% of families and single individuals owe tax. Because the exclusion has been rising, the share of decedents' estates paying estate taxes has been falling. Even prior to the 2001 tax cuts, however, only 2% of decedents paid an estate tax. Progressivity in the tax system, however, could also be altered through changes in the income tax.

Another argument made by proponents of the estate and gift tax is that, to the extent that inherited wealth is seen as windfall to the recipient, such a tax source may be seen by some as fairer than taxing earnings that are the result of work and effort.

Finally, many economists suggest that an important rationale for maintaining an estate tax is the escape of unrealized capital gains from any taxation, since heirs receive a stepped-up basis of assets. Families that accrue large gains through the appreciation of their wealth in assets can, in the absence of an estate tax, largely escape any taxes on these gains by passing on the assets to their heirs. The base of the estate tax is, however, quite different from the base of the capital gains tax, and the rates are higher. The 2001 tax revisions proposed to tax capital gains after allowing a significant exemption.

ARGUMENTS AGAINST THE ESTATE AND GIFT TAX

An important criticism of the estate and gift tax is that it reduces savings and economic growth. However, as is also the case for the income tax, neither economic theory nor empirical evidence clearly indicate that the estate tax reduces savings. For example, while the estate tax may discourage saving for bequests because the cost of making a net bequest (in terms of forgone consumption) increases, the tax also requires a greater amount of saving to achieve a net target. Estate and gift taxes are unlikely to have much effect on assets accumulated for precautionary purposes. Bequests can also reduce saving by heirs because they increase resources for consumption.[2]

A second major argument against the estate and gift tax is that it burdens family businesses and farms and makes it more difficult to pass on these assets to the next generation who can continue the business. However, only a small portion (less than 5%) of businesses and farms are likely to be affected; many of those have sufficient liquid assets to pay the tax. In addition, extensions of time to pay the tax are allowed.[3]

Critics also argue that death is not an appropriate occasion to impose a tax; indeed, the tax is sometimes referred to as a "death tax." Another argument is that wealth has already been taxed through income taxes, though this is not the case for unrealized capital gains. Finally, critics assert that the complexity of the tax not only imposes administration and compliance burdens but undermines the progressivity of the tax.[4] Of course, this latter argument could also be a justification for reforming rather than reducing or abolishing the tax.

OTHER ISSUES

Two other economic effects of the tax that might be considered in evaluating changes are the possible negative effect on charitable contributions (because charitable contributions are deducted from the estate and gift tax base) and the effect on state and local estate taxes. A credit has been allowed against estate and gift taxes for state estate taxes in the past, although this credit has now been eliminated; these changes may create pressure on states to reduce these taxes which will now become a more visible burden on their residents.

REFERENCES

[1]   For a more extensive discussion of estate and gift tax issues, see CRS Report RL30600, *Estate and Gift Taxes: Economic Issues,* by Jane Gravelle and Steven Maguire. See CRS Report RL32818, *Estate Tax Legislation in the 109th Congress*, by Nonna A. Noto for further information on proposals. See also CRS Report 95-416, *Federal Estate, Gift, and Generation-Skipping Taxes: A Description of Current Law*, by John R. Luckey.

[2]   For an analysis of these savings effects and indications of a limited and uncertain effect from empirical data, see William G. Gale and Maria G. Perozek, "Do Estate Taxes Reduce Savings?" and Wojciech Kopczuk and Joel Slemrod, "The Impact of the Estate Tax on Wealth Accumulation and the Avoidance Behavior of Donors," both in *Rethinking Estate and Gift Taxation*, eds. William G. Gale, James R. Hines, Jr. and Joel Slemrod, Washington, D.C., The Brookings Institution, 2001.

[3]   For analyses of the estate tax data, see CRS Report RL30600, *Estate and Gift Taxes: Economic Issues* by Jane G. Gravelle and Steven Maguire; CRS Report RS20593, *Asset Distribution of Taxable Estates: An Analysis*, by Steven Maguire; and Estate Taxes, Life Insurance, and Small Business by Douglas Holtz-Eakin, John W. Phillips, and Harvey S. Rosen, National Bureau of Economic Research working Paper 7360.

[4]   For a discussion of avoidance methods and an estimate that indicates reasonable
       compliance and administrative costs, see Charles Davenport and Jay Soled, "Enlivening
       the Death-Tax Death-Talk." *Tax Notes*, July 26, 1999, pp. 591-629.

INDEX

D

E